Practical Poker Ma...

By Pat Dittmar

D1232878

Published by ECW Press
2120 Queen Street East, Suite 200
Toronto, Ontario, Canada M4E 1E2
416.694.3348 / info@ecwpress.com

LIBRARY AND ARCHIVES CANADA CATALOGUING IN PUBLICATION

Dittmar, Pat
Practical poker math / Pat Dittmar.

ISBN 978-1-55022-833-5

1. Poker. 2. Game theory. I. Title.

GV1251.D48 2008   795.41201'51   C2008-900792-1

Cover Design: Tania Craan
Cover image: BICYCLE®, BEE®
Text Design: Peggy Payne
Typesetting: Gail Nina
Second printing: Marquis

This book is set in Minion.

PRINTED AND BOUND IN CANADA

ECW PRESS
ecwpress.com

Thanks to some Friends for their labors and for what they have taught me

Constantin Dinca
Learned Mathematician and Programming Wizard

Peggy Payne
For Bringing Order to the Chaos

Alice Jue
Lady of Art, Letters and Vodka Martinis

Bernadette Castello
For Everything

About the Author

Today Pat Dittmar is head of trading and technical development at TradePointTechnologies.com. At Trade Point Technologies a small group of poker players use their poker skills and state of the art, proprietary technology to play in the one true Big Game: The World Financial Markets.

The author's math expertise comes from years in the financial community, where he held positions including partner in a registered brokerage firm, derivatives trader, big block trader, arbitrageur and Compliance Registered Options Principal.

As a longtime resident and habitué of Las Vegas, Pat has played professional poker on and off for more than 2 decades and has worked as a paid consultant to a major online poker site.

After a bad beat in the bond market, he went back to sea. Operating under his license as a U.S. Merchant Marine Master Mariner, Pat served as the Captain of small freighters. Much of this work took place along the trading routes off the west coast of Africa.

For many years, in special, protected rooms in Pointe Noire — Old Congo, Port Gentil — Gabon, and Abidjan — Cote d'Ivoire a colorful assortment of ships' Captains, mercenaries, black marketeers, spies and Angolan diamond smugglers gathered to play poker, and on occasion, so did Pat.

Table of Contents

Preface ix

Introduction xi

1. Introduction to Game Theory in Poker 1

    Game Theory – A Historical Perspective 1

    Game Theory and Poker 3

2. The Basic Calculations 9

    Odds & Probabilities 9

    Combinations, Permutations & Factorials 14

    Money Odds & Expectation 19

    Total Odds & Real Expectation 22

    Odds in Texas Hold'em 24

    Odds in Omaha Hi-Lo 34

3. Odds in Texas Hold'em 45

    Starting Hands 49

        Odds & Probabilities 51

        The Calculations 52

        Pocket Pairs 53

        Suited Cards 58

        2 Big Cards 62

        Rags 65

    Before the Flop 68

        The Calculations 71

        Pocket Pairs 71

        Suited Cards 77

        2 Suited in the Hole 78

        2 Big Cards 88

    After the Flop 93

        Odds of Improvement 93

        Money Odds, Overlays & Expectation 93

        From a Game Theory Perspective 95

Odds the Turn Card Will 97
Odds with 2 Cards to Come 103
Odds of Hitting on Either or Both Turn & River 103
Runner – Runner 114
The Calculations 115
Before the Last Card 125
Odds of Improvement 126
The Calculations 126
The River Bet – Last Chance to Earn 133
4. Consolidated Odds Tables 135
Texas Hold'em Odds Tables 136
Starting Hands 136
Before the Flop 136
After the Flop 138
Last Card (The River) 140
Omaha Hi-Lo Odds Tables 141
Starting Hands 141
Before the Flop 141
Odds of Improvement with the Turn Card 143
Odds of Improvement with 2 Cards to Come 143
Odds of Making Certain Runner-Runner Draws 144
Odds of Improvement with Only the River
      Card to Come 145
5. Odds in Omaha Hi-Lo 147
Starting Hands 149
Odds & Probabilities 149
The Calculations 150
Before the Flop 171
Omaha Hi-Lo – Before the Flop 172
The Calculations 173
After the Flop 191
Nut Hand or Nut Draw 191
Money & Expectation After the Flop 191
Odds of Improvement 193
Odds the Turn Card Will 194
The Calculations 195

Odds with 2 Cards to Come 201
    The Calculations 203
    Runner – Runner 215
    The Calculations 215
Before the Last Card 223
    The Calculations 225
The River Bet 231

Preface

Without a clear understanding — or at least a great natural instinct for odds, probabilities and the ratios of risk and reward — success at poker is unlikely.

However, with enough knowledge of these odds and ratios to recognize situations of positive expectation, and the discipline to operate only in an environment of positive expectation, **long-term poker success is virtually assured.**

To get the maximum possible value from every hand, a player need only have perfect knowledge of

1. The cards yet to come

2. The cards held by opponents

3. The future actions and reactions of opponents.

While such perfect knowledge is never available, an understanding of **odds, probabilities and basic game theory** can go a long way toward telling

♦ How likely it is that certain cards will appear or not appear in any given situation

♦ How probable it is that certain opponents might hold certain powerful hands

♦ How opponents are likely to act or react when examined through a Game Theory lens.

This is not perfect knowledge. It is, however, **very powerful poker decision support information.**

Traditionally, pot or implied odds have been compared to a player's odds of improving his hand to calculate expectation. In this book we introduce the concept of a player's **Total Odds** of winning the pot. **Total Odds** includes not only the odds that his hand is or will improve to the best hand, but also the likelihood that the right move at the right time will cause his opponents to fold and give up the pot.

In any given poker situation, a player with a **positive expectation** may **raise** the action, and his opponent(s) will have only 2 choices:

1. **Continue to play at a disadvantage and for higher stakes**

2. **Refuse the raise and forfeit the pot.**

This book presents a practical way to combine the application of traditional poker odds and probabilities with basic game theory to give players at every level of skill and experience a crystal with prisms that will allow them to see the best and the worst of their possibilities, their opponents' strengths and their own best courses of action.

Introduction

Certain practitioners can predict, with perfect accuracy, such natural phenomena as the day, the night, the tides and even celestial events that will occur a thousand years from now.

In poker, any player can predict with that same astonishing accuracy the likelihood of any card appearing at any time. He can predict the likely holdings of his opponents and, based on his own hand, he can predict the long-term profitability of any call, bet or raise.

In astrophysics and the navigation of spacecraft, the requirement for accuracy is absolute. In poker a close approximation is all you need.

The outcome of any hand of poker is determined by either or both of the fall of the cards and the actions of the players. If the cards fall so that you have the best hand and you don't fold, you will win the pot. If you employ a betting strategy that compels your opponent(s) to fold, your cards could be blank and you will still win the pot.

Knowledge of odds and probabilities can turn seemingly random events, such as the fall of the cards, into eminently predictable occurrences. An understanding of positive or negative expectation will tell the long-term profitability of any given play and a grasp of basic Game Theory can tell much about the likely responses of opponents.

The first aim of this book is simplicity and clarity so that any player will be able to access the power of odds, probability and game theory information in support of each poker decision.

To facilitate access, the information in *Practical Poker Math* is organized into layers. For both Texas Hold'em and Omaha Hi-Lo it is presented sequentially, based on the round of betting. For each round there is a brief discussion of applicability, followed by a table of the odds for that round, followed by an expansion and explanation of the calculation of each of the odds found in the table. In the center of the book is a consolidation of all of the odds tables from both games.

The result of this organization is that the player who is only interested in referencing certain odds may easily do so in the chapter that contains the consolidated odds tables — without having to wade through hundreds of odds calculations. The player more interested in the principles may read each section's text and refer to the attendant table. Any player who would like to explore the calculation of a certain set of odds can find the expansion and explanation of that calculation in a logical location by the round of betting. And, finally, for the student who wants to learn it all — it is all there.

1. Introduction to Game Theory in Poker

Game Theory – A Historical Perspective

Some believe that the study of Game Theory began with the works of Daniel Bernoulli. A mathematician born in 1700, Bernoulli is probably best known for his work with the properties and relationships of pressure, density, velocity and fluid flow. Known as "Bernoulli's Principle," this work forms the basis of jet engine production and operation today. Pressured by his father to enter the world of commerce, he is also credited with introducing the concepts of expected utility and diminishing returns. This work in particular can be of use when "pricing" bets or bluffs in no-limit poker.

Others believe the first real mathematical tool to become available to game theorists was "Bayes' Theorem," published posthumously in England in the 18th century. Thomas Bayes was born in 1702 and was an ordained minister. His work involved using probabilities as a basis for logical inference. (The author has developed and used artificially intelligent systems based on "Bayes' Theorem" to trade derivatives in today's financial markets.)

Yet still others believe that the study of Game Theory began with the publication of Antoine Augustin Cournot's *The Recherches* in the early 1800s. The work dealt with the optimization of output as a best dynamic response.

Émile Borel was probably the first to formally define important concepts in the use of strategy in games. Born in 1871 in Saint-Affrique, France, he demonstrated an early penchant for mathematics. In 1909 the Sorbonne created a special chair of "Theory of Functions" which Borel held through 1940. During the years 1921–27 he published several papers on Game Theory and several papers on poker. Important to poker players are his discussions on the concepts of imperfect information, mixed strategies and credibility.

In 1944 Princeton University Press published *Theory of Games and Economic Behavior* by John von Neumann and Oskar Morgenstern. While not the first work to define certain concepts of strategy in games, it is widely recognized as one that has fostered Game Theory as we know it today.

Also important to poker players is the work of Julia Bowman Robinson. Born in 1919, she discovered her passion for mathematics after a bout with scarlet fever and was the first woman admitted to the Academy of Sciences. For poker players her most important work was *An Iterative Method of Solving a Game.*

Credited by many as being a primary shepherd of modern Game Theory is John Forbes Nash, Jr. Diagnosed as a paranoid schizophrenic, Nash was long troubled by delusions. This condition, now in remission, became the subject of a popular film. His work earned him 1/3 of the 1994 Nobel Prize in Economic Sciences. Born in the Appalachian town of Bluefield, West

Virginia in 1928, Nash became well known for a 28-page work he did at age 21 which defined his "Nash Equilibrium" concerning strategic behavior in non-cooperative games. Poker players are most drawn to the story that while he was in a bar near Princeton and being goaded into approaching an attractive blond-haired lady, he suddenly shouted and then ran off to complete his work on "The Mathematics of Competition" which is one basis of Game Theory today.

Game Theory and Poker

More than the study and application of a set of principles, Game Theory in poker is primarily about two goals:

1. The study and understanding of the opposition

2. The development of an efficient strategy to dominate the competition.

To be of maximum value, this study and understanding must be translated into effective action, and these actions must be the antithesis of everything a poker opponent thinks or does.

The difference between an antithesis and a correct response defines the utility of Game Theory in poker. An opponent makes an obvious bluff. You are certain that your hand will not even beat the bluff. A correct response in poker is to fold your hand when you know you are beat. The antithesis is to raise or re-raise and make your opponent fold his.

This is an example of a move. In an environment of ever-increasing odds and stakes such as a poker tournament, good hands just don't come along often enough for a player to make it on the strength of his hand alone. A winning player **must** make moves and the study and application of the principles of Game Theory can help him to know

1. When and where to make the move

2. How likely the move is to succeed.

Expressed in the specific terminology of advanced theorists, poker can be defined as an *asynchronous, non-cooperative, constant-sum (zero-sum), dynamic game of mixed strategies.* While the game is played in an atmosphere of *common knowledge* and no player possesses *complete knowledge*, some players are better able to process this *common knowledge* into a *more complete knowledge* than are their opponents. A player is most able to make the *best-reply dynamic* (sometimes referred to as the *Cournot adjustment*) and earn a *cardinal payoff* after using the process of *backward induction* to construct and deploy a *dominant strategy.*

In poker, beyond a certain set of rules, players do not cooperate with each other because each is trying to win at the expense of all others *(zero-sum).* Moreover, they will repeatedly change their strategies at different intervals and for different reasons (therefore, *asynchronously*).

While information about stakes, pot-size, board-cards and players' actions and reactions is available as *common knowledge* to all players at the table, no player has *complete knowledge* of such factors as the other players' cards or intentions.

The one characteristic common to most outstanding players is their ability to better process and better use — that is, they get more value from — the information that is commonly available to everyone else at the table.

Two of the most basic assumptions of Game Theory are that all players

1. Have equal common knowledge

2. Will act in a rational manner.

But in poker, while all players at any given table have access to the same *common* information, not all of them are smart enough to do something with it. Players who know more about odds and probabilities, and whose instincts and keen observation enable them to better process the common knowledge around them, will take far better advantage of this information and will have correspondingly higher positive expectations.

So while all the players at the table have access to the same *common knowledge*, some players are able to base their actions on knowledge that is *more complete*. That all players in all games will always act rationally is never a safe assumption in poker.

Equilibrium versus Evolution

According to the "Nash Equilibrium," a game is said to be in a state of equilibrium when no player can earn more by a change in strategy. It has been argued that, using the process of backward induction, players will evolve their strategies to the point of equilibrium.

In poker, the astute player's strategy will always be in a state of evolution so that his opponents, in order not to be dominated, will also be compelled to modify their strategies. In a real game of poker there is constant evolution and therefore hardly ever a point of absolute equilibrium.

Efficiency and Diminishing Returns

The concept of *efficiency* and the notion of *diminishing returns* relate to the conservation and the most effective use of power/momentum and resources. In poker this is about a player's chips, cards and table image. These two concepts are particularly helpful to the no-limit player when it comes to "pricing" bets and raises.

In any game, and most especially in no-limit poker tournaments, a player's "Mo" (power) comes from an amalgamation of

1. His image at the table

2. The size of his stack

3. The strength of his hand.

If a player is rich in one, he can lean on the other two. If a player has a very strong hand, not only will he win chips, he will add to his image at the table. If he is possessed of a strong image, he can steal with smaller bets. If his stack is huge, players with smaller stacks will often avoid conflict and he will be able to play weaker hands.

In tournament poker, a player's stack goes down in value with the passage of time. At any point, a player's stack size is finite and, therefore, he must carefully consider pricing when he makes his bet. He must consider

1. At what price will his opponent be "priced out" of the hand

2. At what price will his opponent be "priced into" calling

3. Given a careful study of the opposition, what is the best price to achieve the best result without risking more chips than absolutely necessary.

Conventional poker wisdom has long held that a primary obligation of the big stack is to knock out the small stacks. A game theorist would say that the primary obligation of the **big stack** is to **not pump up** the small stacks. The increasing stakes and other, more desperate players will eliminate most of the short stacks — which will leave the big stacks to carefully pick their spots and eliminate anyone left.

Strategy

The value of applying Game Theory principles in any arena is primarily to help a player develop an efficient strategy that dominates the competition.

From a practical perspective, Game Theory is about strategic development. As no effective strategy is likely to be developed in a vacuum, its formation must be the result of a close study of not only the opposition but also past interactions.

As more players play more poker, many of them increase their skills, and the result is that the game becomes ever tougher to beat.

The players who have dominated the game in the past and the players who will do so in the future are the few who can convert the *common knowledge* available to every player at the table into more *complete knowledge*.

Assuming that all greatly successful players are possessed of advanced knowledge of or instinct for odds and strategy, the primary property of their game that differentiates them from the rest of the field is an uncanny ability to discern an opponent's strength and likely action or reaction.

Today the bulk of poker is about one game — No-Limit Texas Hold'em. In a game where you can either greatly increase or lose all your chips in a single move, knowledge of opponents and their tendencies in certain situations becomes much more important than either stack size or the power of a hand.

With complete information on odds and strategy available to a growing and increasingly more able and competitive pool, the one area of opportunity open to aspiring poker kings is the study and application of Game Theory in poker.

The most difficult calculation in all of poker, especially in No-Limit Texas Hold'em, is that of a player's **Total Odds** of winning the pot. This calculation includes far more than the odds of making a certain hand — it includes the odds of making the hand, the likelihood that your opponent's hand is stronger or weaker, and (most difficult of all) the probability of an opponent's action or reaction to your action or reaction.

The only hope of coming to a reasonably accurate calculation of this matrix of probabilities is via the very essence of Game Theory. You must make a detailed and almost instantaneous analysis of your opponent and his strengths, weaknesses and other propensities.

2. The Basic Calculations

Odds & Probabilities

Odds and Probabilities are simply two methods of expressing the relationship of **Positives**, **Negatives** and **Total Possibilities**. They represent the same basic information and each can be easily converted into the other.

Odds

Odds in poker are expressed as a ratio of the **Negatives** to the **Positives**:

WILLNOTs : WILLs

With a well-shuffled deck of cards, what are the **Odds** that the first card from the deck will be one of the **4** Aces?

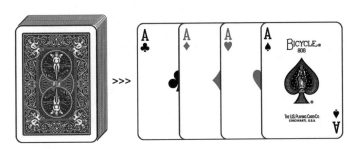

Total Possibilities = 52 Cards in the Deck
Positives/WILLs = 4 Aces in the Deck
Negatives/WILLNOTs = 48 Cards not Aces

To calculate odds or probability, you only need to know any 2 of 3 values: **WILLs**, **WILLNOTs** and **Total Possibilities**. There are 3 easy ways to find the third value:

Total Possible (52) – WILLs (4) = WILLNOTs (48)
Total Possible (52) – WILLNOTs (48) = WILLs (4)
WILLNOTs (48) + WILLs (4) = Total Possible (52)

In a deck of **52** cards (**Total Possibilities**) there are **4** cards (**WILLs**) that are Aces and **48** cards (**WILLNOTs**) that are not Aces.

The odds of drawing an Ace is the ratio of the cards that **WILLNOT** be Aces to the cards that **WILL** be Aces:

WILLNOTs : WILLs
48 : 4
Reduce
48 / 4 : 4 / 4
12 : 1

The odds against drawing an Ace are **12** to **1**.

Probabilities

Probabilities are expressed as either a decimal ratio to 1 or as a percentage. Here is an easy-to-use formula to calculate either:

$$\frac{\textbf{WILLs}}{\textbf{Total Possibilities}} = \textbf{Probability as decimal ratio to 1}$$

$$\frac{\text{WILLs}}{\text{Total Possibilities}} * 100 = \text{Probability as \%}$$

Total Possibilities is the number of cards in the deck, **52**. The number of cards that **WILL**, is the number of Aces in the deck, **4.**

The probability that the first card off a well-shuffled deck will be an Ace is

$$\frac{\text{WILLs}}{\text{Total Possibilities}} = \text{Probability as decimal ratio to 1}$$

$$\frac{4}{52} = .0769$$

$$\frac{\text{WILLs}}{\text{Total Possibilities}} * 100 = \text{Probability as \%}$$

$$\frac{4}{52} * 100 = 7.69\%$$

Another way to calculate probabilities is to multiply the probability of the components. For example, the probability of finding a pair of Aces in the hole in Hold'em can be found by

$$\frac{\text{WILLs}}{\text{Total Possibilities}} = \text{Probability as decimal ratio to 1}$$

With **6** possible pairs of Aces and **1,326** possible starting hands, the probability of Aces is

$$\frac{6}{1,326} = .0045249$$

Another way to get the same answer is to multiply the probability that your first card in the hole will be an Ace by the probability that your second card in the hole will be an Ace:

$$\frac{4}{52} * \frac{3}{51} = .0045249$$

Conversions

Because **ALL** odds and probability calculations begin with two of three basic values (**Total Possibilities, WILLs** and **WILLNOTs**), the conversion of odds to probability or *vice versa* is straightforward.

In the example above the odds are **12 : 1**. Odds are an expression of **WILLNOTs : WILLs**.

To convert these odds to a probability expressed as a percentage:

$$\frac{\textbf{WILLs}}{\textbf{Total Possibilities}} * \textbf{100} = \textbf{Probability as \%}$$

$$\frac{4}{52} * \textbf{100} = \textbf{7.6923\%}$$

To re-convert this **7.6923%** probability back to odds, you know that this probability, **7.6923%**, represents the **WILLs** and that in any probability situation the **Total Possibilities** are **100%.** Thus you have the **WILLs** and only need the **WILLNOTs** to express this probability as odds:

Total Possibilities = 100%
– WILLs = 7.6923%.
WILLNOTs = 92.3077%

WILLNOTs (92.3077) : WILLs (7.6923)
92.3077 : 7.6923
Reduce
92.3077 / 7.6923 : 7.6923 / 7.6923
12 : 1

To complete this circle of conversions, we will reconvert the **12 : 1** odds back to a probability expressed as a decimal ratio to 1.

To calculate the probability expressed as a decimal ratio to 1 you need to know the **WILLs** and the **Total Possibilities**. From odds of **12 : 1** we know that the **WILLs** are **1** and the **WILLNOTs** are **12**. **WILLs(1) + WILLNOTs(12) = Total Possibilities(13)**. You now have the all numbers needed to calculate **12 : 1** as a decimal probability to 1.

$$\frac{\textbf{WILLs}}{\textbf{Total Possibilities}} = \textbf{Probability as decimal ratio to 1}$$

$$\frac{1}{13} = .076923$$

Quickie conversion *finale*: the conversion of the probability of **.076923** back into odds is

Total Possibilities = 1
− WILLs = .076923
WILLNOTs = .923077

WILLNOTs (.923077) : WILLs (.076923)
.923077 : .076923
Reduce
.923077 / .076923 : .076923 / .076923
12 : 1

Combinations, Permutations & Factorials

Whether it is called combinatorics or combinatorial math, it is a branch of discrete mathematics that for our purposes deals with the determination of the size of certain sets.

Most odds and probability calculations can be reduced to 2 basic steps:

1. **Define the Sets** — The calculation of any set of odds or probabilities involves the comparison of 2 of 3 sets: **Total Possibilities**, **WILLs** and **WILLNOTs**.

2. **Compare the Sets** — Odds in poker are an expression of the comparison of the number of **WILLNOTs : WILLs**. Probabilities are an expression of **WILLs / Total Possibilities**.

Formulae that calculate the populations of combinations, permutations or factorials are used to calculate the population of the **WILL** and **WILLNOT** sets.

Mathematicians express combinations as "C(52, 2)," in *Practical Poker Math*, we have used "**Comb(52, 2)**." Permutations are expressed as "P(52, 2)" and we have used "**Perm(52, 2)**." For factorials we have used the standard **52!**.

Combinations

The mechanics of the math used to determine the enumeration of combinations is more easily demonstrated than explained.

Of the 3 (combinations, permutations and factorials) the

most used calculation in poker is the determination of the size of certain combinations.

The calculation of a combination in the format of **Comb(52, 2)**, will produce the number of different 2-card combinations — starting hands — that are possible from a 52-card deck:

$$\frac{(52 * 51)}{(1 * 2)} = 1{,}326$$

To find the number of 3-card Flops possible from the 50 remaining unseen cards after you have been dealt your 2 hole cards in Hold'em, use **Comb(50, 3)**:

$$\frac{(50 * 49 * 48)}{(1 * 2 * 3)} = 19{,}600$$

There are 6 different pairs or 2-card combinations of Aces that can be made from the 4 Aces in a 52-card deck:

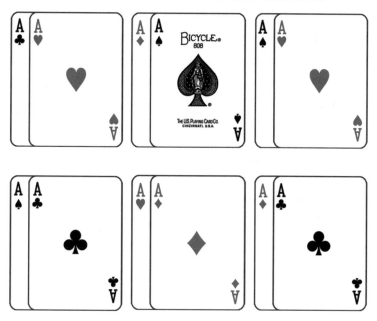

This can be calculated using **Comb(4, 2)**:

$$\frac{(4 * 3)}{(1 * 2)} = 6$$

With 10 great picks for a weekend of sports betting, to find the number of 4-bet parlays — **Comb(10, 4)**:

$$\frac{(10 * 9 * 8 * 7)}{(1 * 2 * 3 * 4)} = 210$$

There are basically two ways to calculate probabilities:

1. The comparison of sets

2. The mathematical comparison of probabilities.

The comparison of sets is more commonly used and is helpful when trying to determine the size of multiple sets.

The Lottery

We can use these same formulae to find the odds or probability of winning a state lottery. As an example, use a lottery where, to win, a player must correctly pick 5 numbers in the range of **1** through **47** and **1** correct "special" number from **1** through **27**.

To find the number of possible 5-number combinations in **1** through **47** — **Comb(47, 5)**:

$$\frac{(47 * 46 * 45 * 44 * 43)}{(1 * 2 * 3 * 4 * 5)} = 1,533,939$$

To find the total number of possible **6**-number combinations in **1** through **27**:

$$1,533,939 * 27 = 41,416,353$$

41,416,352 combinations **WILLNOT** win the lottery and only **1 WILL**.

WILLNOTs : WILLs

41,416,352 : 1

Assume that

1. yours was the only winning ticket

2. you took a cash-settlement for half the amount and

3 you have to pay 1/3 of your win in taxes.

To have a positive expectation from the purchase of a $1.00 ticket, the lottery's first prize would have to be on the order of $125,000,000.00.

Permutations

To demonstrate the difference between combinations and permutations, the two hands below are 2 different permutations of the same combination.

The number of 2-card permutations possible from a 52-card deck — **Perm(52, 2)** is:

$$52 * 51 = 2,652$$

At the racetrack, with an 8-horse field, the number of possible trifecta tickets — **Perm(8, 3)** is:

$$8 * 7 * 6 = 336$$

Factorials

Factorials are seldom, if ever, used in poker and are mostly about orderings. A deck of 52 cards contains one combination and many permutations or orderings.

The factorial of 52 is expressed as **52!** (sometimes called 52 bang). On a scientific calculator one would enter 52 and then press the **n!** key.

52! yields the number of possible deck sequences or orderings that can be created from one 52-card deck. The mechanics of the calculation **52!** are

$$52 * 51 * 50 * 49 * 48 *...\text{ through } * 4 * 3 * 2 * 1$$

When you enter **52!** into a scientific calculator the result is **8.065817517 +e67** which (give or take a zero or two) is equal to:

8,065,817,517 plus 58 more zeros

This is the number of possible decks or orderings that may be created from a single deck of 52 cards.

Money Odds & Expectation

The simplest explanation we have for the complex subject of money odds and expectation is

**The Odds will tell you the Risk and
The Pot will tell the Reward**

Put another way: if the money odds are greater than the overall odds of winning the pot, you have a positive expectation.

Pot Odds

Pot Odds are an expression of what the pot is offering on a particular round and doesn't take into consideration any action that might come on future rounds.

Pot Odds reveal nothing about a player's chances of winning a pot, but only indicate the money leverage he is being offered at that moment if he stays in the pot. Money odds are an expression of the ratio of the money in the pot, including previous bets and calls on this round, to the amount it will cost to stay in the hand.

With $600 in the pot plus a $100 bet to you and $100 to call, the pot is offering Pot Odds of **7 : 1**.

Implied Odds

Implied odds are the money odds offered by the current pot plus a best estimate of the money that will be added to the pot on future rounds of betting.

After the Flop, and given a certain draw, a consideration of pot odds alone might indicate a negative expectation for that particular draw.

However, especially in active games, an estimate of the pot size at the river could well offer money odds enough to justify the draw.

Expectation

Expectation is the return that can be anticipated from any given wager or proposition after enough bets have been made to normalize the result.

All experienced poker players are intimately acquainted with the concepts of both positive and negative fluctuation. Results from any given proposition may vary greatly from a correctly calculated expectation in the short term, but will completely normalize themselves after a sufficient sampling.

We know that the odds against drawing an Ace from the top of a well and fairly shuffled deck are **12 : 1** against.

Suppose that you are offered a proposition where you will lose $100 every time the first card off is not an Ace but you will be paid **15 : 1** every time the card is an Ace.

While you will lose 12 of 13 tries under the terms of this proposition, you still have a nice positive expectation and the only 2 things that can stop you from winning a large long-term profit are

1. You stop taking the bet

2. You don't have the bankroll to withstand the inevitable fluctuations.

On average, over the course of 13 such bets you will lose 12 for a loss of $1,200 and win 1 worth $1,500, for a net profit of $300 over 13 tries.

12 losses @ $100 ea = $1,200 loss
1 Win @ $1,500 = $1,500 profit
Net Profit = $300

With a profit of **$300** over 13 tries, you can anticipate an average profit of $300 / 13 = **$23.08 per trial**. The value or expectation of each draw is +**$23.08**.

The sum of all expectations in any proposition must be zero.

You, as the player, have a positive expectation of +**$23.08** per draw and the person who laid the bet has a negative expectation of -**$23.08**.

Fluctuation is short-term.

After 20 draws it would not be unusual for you, as the taker of such a wager, to be **down $2,000** and your opponent, who made the proposition, **up $2,000** — **a fluctuation**. While after a small number of tries you might be down, it is **almost a 100% certainty** that after 100,000 tries you will be **up over $2,000,000** and your opponent will be **down over $2,000,000** — **an expectation**.

Expectation is long-term.

The same applies to every bet or raise you make or take playing poker. As long as all of your bets, calls and raises carry greater money odds than risk odds, you will be a long-term winner.

Total Odds & Real Expectation

Your **Real Expectation** is the **Total Odds** of your winning the pot compared to the ratio of the final value of the pot to what it will cost to implement the optimal strategy to take the pot down.

Your **Total Odds** of winning the pot are the combination of the likelihood of your hand being or improving to the best hand, **and** the probability that the right move at the right time will cause your opponent(s) to fold and forfeit the pot.

A weakness in the **traditional** application of both pot odds and implied odds is that they are usually compared only to your odds of improvement. What is missing is the likelihood that you can also win the pot by making the right "move" even if you fail to make your hand.

For example, assume you are in a game where you have good knowledge of your opponents, you are on the button before the last card, and you have a 25% or 1 in 4 chance of completing a draw to what will surely be the best hand.

A 25% probability of drawing the winning hand is equal to odds of **3 : 1** against. With those odds against, a player must have money odds of **4 : 1** or better to justify a call.

In this hand there are two opponents still involved in the pot. As a student of game theory, you've tracked these two opponents and calculate that there is a 50% chance they will both fold to a bet or raise.

If you only consider your odds of improvement, you have a

25% chance of winning and a 75% chance of losing. Thus, you need money odds of something considerably more than 3 : 1 to justify further participation in the hand.

On the other hand, if you consider your 25% chance of making the draw to win the pot *plus* the 50% chance that your opponents will fold to a bet or raise, your **total** odds of winning the pot are approximately 60%. With a 60% chance of winning, you can participate with money odds of 2 : 1 (or even slightly less) instead of the 4 : 1 that are required if you only consider your 25% probability of improving the hand.

While it is not always possible to accurately predict an opponent's hand or future actions, knowledge of basic odds and player history will give you a strong indication.

If you only consider the odds of improvement, you can expect to win only 25% of the time. But with knowledge of your opponents (and even with some overlap of possibilities), in this situation you can expect to win well over 50% of the time with a bet or raise.

The positive application of this concept of **Total Odds** will help you discover value plays with good positive expectation that you might have otherwise missed.

While it is certainly harmful to your bottom line to play with an underlay, it is equally harmful to your long-term success to pass on value opportunities that offer significant positive expectation.

Odds in Texas Hold'em

Most calculations in Chapter 4 — *Odds in Texas Hold'em* — involve the determination of the odds or probability that a hand will improve.

All of these calculations are made in one of two ways:

1. The comparison of 2 of 3 sets: **Total Possibilities**, **WILLs** and **WILLNOTs**

2. The combination of probabilities.

These methods are fully demonstrated in this chapter and throughout the book.

Below are some sample calculations relevant to each round of betting in Texas Hold'em. A more complete selection of odds and a detailed discussion of these calculations can be found in Chapter 4 — *Odds in Texas Hold'em*.

Starting Hands

Calculating the odds or probability that you will receive any given starting hand involves knowing

- ♦ the number of 2-card combinations that **WILL** be the hand

- ♦ the number of 2-card combinations that **WILLNOT** be the hand

- ♦ the number of **Total Possible** 2-card combinations from a 52-card deck.

To find the number of 2-card combinations possible from a 52-card deck, as demonstrated previously, we will use **Comb(52, 2)**:

$$\frac{(52 * 51)}{(1 * 2)} = 1{,}326$$

The number of pairs possible from any of the 13 ranks in a deck of cards can be calculated by **Comb(4, 2)**:

$$\frac{(4 * 3)}{(1 * 2)} = 6$$

There are 6 possible pairs of Aces:

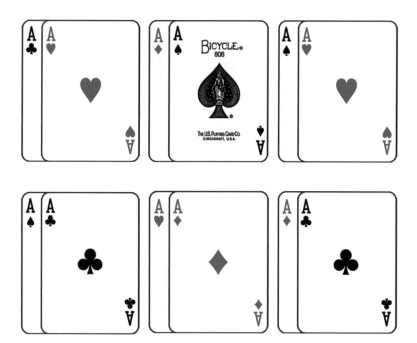

To find the odds of a pair of Aces in the hole, calculate the number of 2-card combinations that **WILLNOT** contain a pair of Aces:

Total Possible (1,326) – WILLs (6) = WILLNOTs (1,320)

WILLNOTs (1,320) : WILLs (6)
1,320 : 6
Reduce
1,320 / 6 : 6 / 6
220 : 1

Another way to find the probability of pocket rockets is to multiply the probability that the first card will be an Ace (**4 / 52**) by the probability that the second card will be an Ace (**3 / 51**):

$$\frac{4}{52} * \frac{3}{51} = .0045249$$

To calculate the odds that you will find ANY pair in the pocket:

There are **6** possible pairs from each of **13** ranks in the deck, so there are **6 * 13 = 78** possible pairs in a deck.

Total Possibilities = 1,326
– WILLs = 78
WILLNOTs = 1,248

WILLNOTs (1,248) : WILLs (78)
1,248 : 78
Reduce
1,248 / 78 : 78 / 78
16 : 1

Before the Flop

Before the Flop, what you want to know most is the odds or probability that the Flop will improve your hand.

The Flop is a set of 3 cards. To calculate the odds or probability that the Flop will make the desired improvement, you will need to know at least 2 of the following 3 values:

1. The total number of 3-card Flops possible from the deck, minus your 2 hole cards

2. The total number of 3-card combinations that **WILL** make the desired improvement

3. The total number of 3-card combinations that **WILLNOT** make the desired improvement.

With 2 cards in the hole there are 50 unseen cards. To find the total number of 3-card Flops possible from the rest of the deck, use **Comb(50, 3)**:

$$\frac{(50 * 49 * 48)}{(1 * 2 * 3)} = 19,600$$

To find the number of 3-card combinations that **WILL** improve your hand depends on the hand.

To find the number of 3-card combinations that **WILLNOT** make the improvement, use **Total Possible – WILLs = WILLNOTs**.

Pocket Pair >>> Set or Full House

To find the odds of flopping a Set or a Full House with a Pair in the pocket, find the number of 2-card combinations possible from the remaining unseen 50 cards minus the **2** cards that will make the Set — **Comb(48, 2)**:

$$\frac{(48 * 47)}{(1 * 2)} = 1,128$$

Multiply this number (**1,128**) by the number of remaining cards that **WILL** make the Set (**2**) to find the number of possible 3-card Flops that will make the Set or a Full House:

$$1,128 * 2 = 2,256$$

There is a possibility that a Set could come on the Flop. This would also make a Full House with the Pair in the pocket.

From the 12 ranks that do not include the rank of the pocket pair, there are 48 different 3-card combinations that are Sets — **Comb(4, 3) * 12**:

$$\frac{(4 * 3 * 2)}{(1 * 2 * 3)} * 12 = 48$$

In all, there are **2,256 + 48 = 2,304** possible 3-card Flops that will make a Set or Full House when holding a pocket Pair.

You now have all the numbers needed to calculate the odds of flopping a Set or a Full House:

Total Possibilities = 19,600
– WILLs = 2,304
WILLNOTs = 17,296

Odds of Flopping a Set or Boat with a Pocket Pair
WILLNOTs : WILLs

17,296 : 2,304
Reduce
17,296 / 2,304 : 2,304 / 2,304
7.5 : 1

Chapters 4 — *Odds in Texas Hold'em* and 6 — *Odds in Omaha Hi-Lo* — explain the odds of flopping Straight and Flush draws, Straights and Flushes, and more.

After the Flop

Once the Flop has hit the board, you have 5 of the 7 cards that will make your best and final hand. After the Flop and with 2 cards to come, regardless of the hand, any player still involved must know

1. The odds of improvement on the next card (the Turn)

2. The odds of improvement over the next two cards (the Turn and the River)

3. The odds of hitting certain Runner-Runner hands.

Odds of Improvement on the Turn

After the Flop, the calculation of the odds of improvement with the next card is straightforward. The only number that changes is the number of outs.

After the Flop and before the Turn, counting your 2 hole cards and the 3 Flop cards, there are **47** unseen cards.

Open Ended Straight Draw >>> Straight

The odds that an open ended Straight draw will improve into a Straight with the Turn card are easily calculated.

Regardless of how the 4 cards making up the draw lie, there are **8** outs to make the Straight:

Total Possibilities = 47
– WILLs = 8
WILLNOTs = 39

Odds of Turning a Straight with an Open Ended Draw
WILLNOTs : WILLs
39 : 8
Reduce
39 / 8 : 8 / 8
4.9 : 1

Odds of Improvement on Either/Both the Turn or the River

After the Flop and before the Turn card has been dealt, there are **47** unseen cards. When calculating the odds with **2** cards to come, one important number is the total number of possible 2-card combinations that can be made from the remaining unseen **47** cards — **Comb(47, 2)**:

$$\frac{(47 * 46)}{(1 * 2)} = 1,081$$

With 2 cards to come there are 4 possibilities:

    1. You will hit none of your outs on either card

2. You will hit one of your outs on the Turn

3. You will hit one of your outs on the River

4. You will hit one of your outs on the Turn and another on the River.

Open Ended Straight Draw >>> Straight with 2 Cards to Come

The odds of completing an open-ended Straight draw with 2 draws to come are a lot better than with only one draw left. In the above example, the odds of making the Straight on the Turn are almost **5 : 1**. As you can see below, the odds of making the Straight on EITHER or both the Turn or River are just over **2 : 1**.

Of 47 unseen cards, **8** will make the straight and **39** will not:

$$8 * 39 = 312$$

312 2-card combinations will contain 1 and only 1 Straight card.

$$\frac{(8 * 7)}{(1 * 2)} = 28$$

28 2-card combinations will contain 2 Straight cards including the possibility of a Pair.

312 + 28 = 340 possible 2-card combinations that **WILL** make the Straight:

Total Possibilities = 1,081
– WILLs = 340
WILLNOTs = 741

**Odds of a Straight with Open Ended Draw
and 2 Cards to Come
WILLNOTs : WILLs
741 : 340
Reduce
741 / 340 : 340 / 340
2.18 : 1**

Runner – Runner

After the Flop and with 2 cards to come, you will need help on both of them. **No one card will do it.**

Remember — runner-runners don't complete very often and in most cases require extraordinary money odds to be played profitably.

To prove this point, let's consider the odds:

♦ To make the Straight with 2 cards to come, the odds are a little over **2 : 1**

♦ With only 1 card to come they are nearly **5 : 1**

♦ When runner-runner is required they are over **20 : 1** against.

Open Backdoor 3-Straight >>> Straight

With 3 connected cards above **2** and below **K** in rank, there are 3 sets of 2-card combinations that will make the Straight. For each of these **3** sets, there are **4 ∗ 4 = 16** possible combinations. Therefore, there are a total of **3 ∗ 16 = 48** 2-card combinations that will make the Straight:

Total Possibilities = 1,081
– WILLs = 48
WILLNOTs = 1,033

Odds of a Runner-Runner Straight
WILLNOTs : WILLs
1,033 : 48
Reduce
1,033 / 48 : 48 / 48
21.52 : 1

Before the Last Card

With one card to come, the odds calculation for the various hands is straightforward. Here again, the only number that changes is the number of outs.

After the Turn card has hit the board, there are **2** cards in your pocket, **4** on the board and **46** unseen cards.

Total Possibilities = 46

Total Possibilities – WILLs (Outs) = WILLNOTs

Odds = WILLNOTs : WILLs (Outs)

There are **8** cards (4 on each end) that will make the Straight:

Total Possibilities = 46
– WILLs = 8
WILLNOTs = 38

Odds of a Straight with an Open Ended Draw
WILLNOTs : WILLs
38 : 8
Reduce
38 / 8 : 8 / 8
4.75 : 1

The odds of making the Straight on the River are slightly better than the odds of making the same Straight on the Turn, because there is one less card that **WILLNOT** make the Straight among the unseen cards.

Odds in Omaha Hi-Lo

A major difference between calculating the odds in Texas Hold'em and the odds in Omaha Hi-Lo is that many Hi-Lo hands have opportunities to win in two ways.

Another difference is with starting hands. A player is more

than five times as likely to see Aces in the hole in Omaha as he is in Hold'em.

Omaha Hi-Lo is one of the few games in poker where the absolute nut hand can wind up losing money. It is also a game where, given usually large pots, playing unidirectional hands (regardless of strength) can carry a much-reduced expectation over the same hand in non-split games.

In split games, the difference between scoop and not-scoop is dramatic. If 4 players each have $200 invested in an $800 Omaha Hi-Lo pot:

Scooper Earns $600 Profit
Low Half Earns $200 Profit
Quartered Low Earns $000 Profit

Starting Hands in Omaha Hi-Lo

A number that will prove useful throughout these calculations is the total number of possible 4-card combinations in a 52-card deck — **Comb(52, 4)**:

$$\frac{(52 * 51 * 50 * 49)}{(1 * 2 * 3 * 4)} = 270{,}725$$

There are **270,725** different starting hands possible in Omaha.

The Odds of AA in the Hole in Omaha

The odds of getting a pair of Aces in the hole in Omaha Hi-Lo underscores the dramatic difference between Hold'em (with 2 cards in the pocket) and Omaha Hi-Lo (with 4).

The odds of pocket rockets in Hold'em are **220 : 1** and the odds of **AAxx** in the hole in Omaha Hi-Lo are **39 : 1**, so you will see Aces in the hole more than 5 times as often in Omaha Hi-Lo as you will in Hold'em.

These odds can be calculated for exactly 2 Aces in the hole or for 2 or more Aces in the hole. The calculation that will serve you best is for exactly 2 Aces in the hole, because in Omaha, having 3 or more cards of the same rank in the hole is NOT a good thing. Omaha is one of the very few places in all of poker where a hand containing 4 hidden Aces is a huge disappointment.

Start with **270,725** — the total possible 4-card combinations in a deck of 52 cards.

Find the number of 2-card combinations that contain a pair of Aces — **Comb(4, 2)**:

$$\frac{(4 * 3)}{(1 * 2)} = 6$$

The last number in this series is the number of 2-card combinations that contain no Aces — **Comb(48, 2)**:

$$\frac{(48 * 47)}{(1 * 2)} = 1,128$$

To find the total number of 4-card combinations that contain **exactly** 2 Aces, multiply the number of 2-card combinations that **WILL** contain 2 Aces (6) by the number of 2-card combinations that **WILLNOT** contain any Aces (1,128):

$$6 * 1,128 = 6,768$$

The probability of 2 and only 2 Aces in the pocket in Omaha Hi-Lo expressed as a percentage is:

$$\frac{\textbf{WILLs}}{\textbf{Total Possibilities}} * 100 = \textbf{Probability as \%}$$

$$\frac{6.768}{270,725} * 100 = 2.5\%$$

To find the odds of a pair of Aces in the hole in Omaha Hi-Lo:

Total Possibilities = 270,725
– WILLs = 6,768
WILLNOTs = 263,957

WILLNOTs (263,957) : WILLs (6,768)
263,957 : 6,768
Reduce
263,957 / 6,768 : 6,768 / 6,768
39 : 1

Before the Flop

The total number of possible different 3-card Flops will be used throughout these calculations. With 4 hole cards, to determine the total possible **3**-card Flops from the remaining **48** unseen cards — **Comb(48, 3)**:

$$\frac{(48 * 47 * 46)}{(1 * 2 * 3)} = 17{,}296$$

Another number that will be useful in these calculations is the number of **2**-card combinations possible from the remaining unseen **48** cards — **Comb(48,2)**:

$$\frac{(48 * 47)}{(1 * 2)} = 1{,}128$$

AA2X >>> Nut Low or Aces Full

With a premium starting hand and 2-way possibilities, you have excellent odds of making a very good hand in either or both directions.

To flop a Nut Low with this hand, the board must contain 3 unpaired cards ranked **3** through **8**. There are 6 ranks of 4 suits, making **Comb(24, 2)** 2-card combinations:

$$\frac{(24 * 23)}{(1 * 2)} = 276$$

Minus **36** possible pairs among these ranks, there are **276 – 36 = 240** 2-card combinations that will flop a Nut Low draw. Multiply this by all of the remaining cards ranked **3** through **8** that are not represented by the first 2 cards of the Flop (**16**), to produce the number of 3-card combinations that will make a Nut Low to this hand:

$$240 * 16 = 3{,}840$$

To flop Aces Full the board must contain an Ace and any pair. With this hand there are **66** unseen pairs ranked **3** through **K**, plus **1** pair of Aces and **3** pairs of Deuces, for a total of **70** possible pairs that might appear on the board. Multiply this by the number of unseen Aces (**2**), to produce the number of possible 3-card Flops that will make Aces Full or better to this hand:

$$70 * 2 = 140$$

This calculation includes the possibility of flopping a pair of Aces to produce QUADS.

There are **3,840 + 140 = 3,980** possible 3-card Flops that will make a Nut Low hand or Aces Full or better to this hand. The odds of this Flop are:

Total Possibilities = 17,296
– WILLs = 3,980
WILLNOTs = 13,316

Odds of Flopping a Nut Low or Aces Full
WILLNOTs : WILLs
13,316 : 3,980
Reduce
13,316 / 3,980 : 3,980 / 3,980
3.3 : 1

After the Flop

Once the Flop has hit the board, you have **7** of the **9** cards that will make your best and final hand for both low and high. Money expectations aside, as a general rule, **only continue to invest in the hand with the nuts or a reasonable draw to the nuts in either direction, preferably both.** If you don't have them, the nuts that is, there are many chances that at least one of your opponents does.

After the Flop and with 2 cards to come, regardless of the hand, any player still involved must know:

1. The odds of improvement on the next card (the Turn)

2. The odds of improvement over the next two cards (the Turn and the River)

3. The odds of hitting certain Runner-Runner hands.

Odds of Improvement on the Turn

With the Turn card to come, the odds calculation for the various hands is straightforward. The only number that changes is the number of outs.

After the Flop and before the Turn, counting your 4 hole cards, there are **45** unseen cards.

Low Draw & Flush Draw >>> Low or Flush

There are **16** cards that will complete the Low, plus **9** cards that will make the Flush, minus **4** cards that are counted among those that will make the Low — **16 + 9 − 4 = 21**:

Total Possibilities = 45
− WILLs = 21
WILLNOTs = 24

Odds of Turning a Low or a Flush
WILLNOTs : WILLs
24 : 21
Reduce
24 / 21 : 21 / 21
1.1 : 1

Odds With 2 Cards to Come

With 2 cards to come there are 4 possibilities:

1. You will hit none of your outs on either card

2. You will hit one of your outs on the Turn

3. You will hit one of your outs on the River

4. You will hit one of your outs on the Turn and another on the River.

When calculating the odds with **2** cards to come, one important number is the total number of possible 2-card combinations that can be made from the remaining unseen **45** cards — **Comb(45, 2)**:

$$\frac{(45 * 44)}{(1 * 2)} = 990$$

To find the **WILLs**, add the number of possible 2-card combinations that will contain **1** and only 1 of the available outs to the number of 2-card combinations that will contain **2** out cards.

Low Draw >>> Low with 2 Cards to Come

With a Low draw, **16** of the unseen 45 cards will make a Low. There are also **6** cards among the 45 that will counterfeit one of the low cards in the hole. Of the 45 unseen cards, there are **16** that will make the Low and **23** that can combine with one of these 16 to make the Low without counterfeiting one of the low cards in the hole.

Therefore, **16 * 23 = 368** is the number of 2-card combinations, one card of which will make the Low.

Comb(16, 2) is the number of 2-card combinations that contain 2 cards that will make the Low, including pairs:

$$\frac{(16 * 15)}{(1 * 2)} = 120$$

120 2-card combinations will contain 2 low cards and still complete the Low without counterfeiting either of the low cards in the hole.

368 + 120 = 488 possible 2-card combinations that **WILL** make the Low:

Total Possibilities = 990
– WILLs = 488
WILLNOTs = 502

Odds of a Low with 2 Cards to Come
WILLNOTs : WILLs
502 : 488
Reduce
502 / 488 : 488 / 488
1.03 : 1

$$\frac{\textbf{WILLs}}{\textbf{Total Possibilities}} * 100 = \textbf{Probability as \%}$$

$$\frac{488}{990} * 100 = 49\%$$

Odds of Runner – Runner

For runner-runner analysis, one number we need is the total possible 2-card combinations from the remaining **45** unseen cards:

$$\frac{(45 * 44)}{(1 * 2)} = 990$$

Back Door Low Draw >>> Runner-Runner Low

When playing Omaha Hi-Lo, it is often of great interest to know the odds of making a backdoor Low.

With 2 low cards in the hole and 1 usable low card on the board, even if the board low card is paired, there are **20** cards of 5 ranks, any unpaired 2 of which will make the runner Low — **Comb(20, 2)**, minus the **30** pairs possible among the cards from these 5 ranks. Therefore:

$$\frac{(20 * 19)}{(1 * 2)} = 190$$

190 – 30 = 160 possible 2-card combinations **WILL** make a runner-runner Low:

Total Possibilities = 990
– WILLs = 160
WILLNOTs = 830

Odds of a Runner-Runner Low
WILLNOTs : WILLs
830 : 160
Reduce

$$830 / 160 : 160 / 160$$
$$5.2 : 1$$

Before the Last Card

With one card to come, the odds calculation for the various hands is straightforward. The only number that changes is the number of outs.

After the Turn card has hit the board, there are 4 cards in your pocket, 4 on the board and **44** unseen cards.

Low Draw >>> Low

With a Low draw, there are **16** cards from 4 ranks that can come on the River to make the Low:

Total Possibilities = 44
− WILLs = 16
WILLNOTs = 28

Odds of Making a Low
WILLNOTs : WILLs
28 : 16
Reduce
28 / 16 : 16 / 16
1.75 : 1

$$\frac{\textbf{WILLs}}{\textbf{Total Possibilities}} * 100 = \textbf{Probability as \%}$$

$$\frac{16}{44} * 100 = 36\%$$

3. Odds in Texas Hold'em

Texas Hold'em is a magnificent game of strategy and chance. Like the personalities, aspirations and abilities of the players, each hand is different.

Sometimes the only path to victory is to hold the best hand. And sometimes a player is able to execute strategies that will allow him to win even if both of his hole cards are blank.

Over time a player will receive the same cards and opportunities as any other player. Each player at the table has access to the same information as all other players at the table. The degree to which a player will triumph over the table depends on his ability to

1. Process and weigh available information as inputs into his decision support system

2. Devise and implement strategies based on that information.

Hold'em is the one poker variation that rewards skill and understanding above all others.

Basic poker odds and probability processing will tell the astute player some measure of

♦ The relative strength of his hand

♦ The likelihood that his hand will improve

♦ Some idea of what his opponents are holding.

Simple observations will give the astute player a reasonable idea of the likely actions of opponents in certain situations. The poker artiste will use this information plus his advanced gamesmanship to execute strategies that will give him a much higher expectation.

The best poker decisions are those based on the best strategy and the best strategy is based on the best information.

Game theory teaches that strategic thought is mostly about the anticipation of the actions and reactions of your opponents.

Basic odds and probabilities tell a player much about his own hand and its likelihood of success. Those same odds apply individually and collectively to all other players at the table.

It is around the margins of strong hands where a game is optimized. The degree to which a player is able to success-fully model an opponent's likely strength and — from past observations — is able to predict the opponent's likely response is equal to the degree to which he will be able to get optimal return from almost any hand he chooses to play.

The key to enhancing the return from any player's game is learning to weigh all available information to profitably play an otherwise marginal hand.

Adapted for poker, various translations of the writings of *Sun Tzu* have loosely stated that, if in every situation of battle a player

♦ knows more about his own strengths, weaknesses and intentions than is known by his opponents;

♦ knows as much about his opponents' strengths, weaknesses and intentions as is known by his opponents themselves; and

♦ has the *Tao*, in the case of poker — the odds/expectation, on his side;

then, that player will triumph in battle after battle.

While a study of the **Odds** can do much to help you predict the **Cards** to come, it is from a study of your **Opponents** that you can surmise much about the **Action** to come.

All players get the same number of good hands. It is the rare and, oftentimes, extremely successful player who can win pots with very strong hands and very weak hands alike.

The most valuable ability in all of poker is the ability to win <u>without</u> having the best hand.

This chapter is organized to present tables with odds for each round of betting, and demonstrate below each table the calculations of those odds.

Basic calculations will explain how often the player will see certain starting hands, how often those starting hands will complete into winning hands and whether certain hands are worth drawing to from a money/expectation perspective.

Those same basic calculations speak volumes about the likely strength of the opponents. When combined with the knowledge of how a given opponent has behaved in the past, the *artiste* can anticipate his opponent's strength and likely action or reaction.

A player's **Total Odds** of winning the pot combine his odds that he has or will have the best hand with his odds of being able to manipulate his opponents **out** of the hand.

Starting Hands

Your first chance to gain an edge over your opponents is with your starting hands. Some hands, like suited connectors, have longer odds of improving and thus do better when you have more money/players in the pot. Other hands, like big Pocket Pairs such as **AA** or **KK**, are already "**made**" hands and do better with fewer players drawing against them.

Just knowing what hands to play from what position and how to play them will give a player a big edge.

An even bigger edge comes from knowing your opponents and their likely actions and reactions.

This section will give you a good idea of what to expect from starting hands and how often to expect them.

Your Odds = Your Opponent's Odds

Your odds of getting any of the hands shown below are exactly the same as for any one of your opponents.

While it is true that you will find rags in the hole four out of five times — it is also true that four out of five of your opponents will also find rags in the hole.

The tables, calculations and methods below demonstrate the characteristics of a normal distribution of starting hands.

From this, a player can always have an idea of what to expect and how often to expect it, plus a good idea of both the individual and collective expectations of his opponents.

In a full 10-handed Hold'em game, some reasonable expectations about starting hands and some derived points of inference from those expectations are as follows:

♦ Once in an average of approximately every 55 hands (5.5 rounds) seen by you or by any one of your opponents, you or that specific opponent will receive a pair of Jacks or better in the hole. The corollary follows.

♦ Once in roughly every 5.5 hands (slightly less often than twice every round) someone — anyone at the table — will receive a pair of Jacks or better in the hole.

♦ Playing online at 4 tables simultaneously with an average of 60 hands per hour per table, someone at one of the 4 tables will have a pair of Jacks or better in the hole roughly every 80 seconds.

♦ Slightly more often than once every other round (1 in 17 hands) you or any specific one of your opponents will receive a Pocket Pair.

♦ On average, for every 17 sets of hole cards dealt, which is slightly more than once every other hand, someone at the table will hold a Pocket Pair.

♦ With over 80% of possible starting hands designated as rags, with not every non-rag hand playable in every situation, and with the **exception** of **blinds** and **special situations, UPWARDS OF 90% OF ANY PLAYER'S STARTING HANDS SHOULD BE DEEMED UNPLAYABLE.**

♦ On average there will be almost 2 non-rag hands dealt per round. If 3 or more non-blind players pay to see the Flop on a consistent basis, you can tell how loose or tight that game is.

♦ If any particular player pays to see the Flop more than once every three hands, assume that for at least some of those hands, he's playing rags.

Odds & Probabilities

First you will be dealt one of **1,326** possible starting hands. This number results from the combinations formula demonstrated in Chapter 1 — *The Basic Calculations* — **Comb(52, 2)**:

$$\frac{(52 * 51)}{(1 * 2)} = 1,326$$

For purposes of classification, almost all of these **1,326** hands can be sorted into one of four categories:

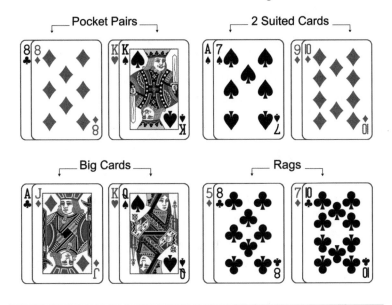

The table below displays both the odds and the probability that any given player will receive certain hands as his first two cards. The calculations to determine these odds and probabilities are demonstrated below the table.

Starting Hands – Odds/Probabilities

Hand	Probability	Odds
Any Pair	5.9%	16 : 1
Any Specific Pair (AA)	.452%	220 : 1
Pair of Jacks or Better	1.8%	54.25 : 1
22 through TT	4.10%	23.6 : 1
Suited Connectors	2.10%	46.4 : 1
Suited Ace	3.62%	26.6 : 1
Any 2 Suited	23.53%	3.25 : 1
Any AK	1.2%	81.9 : 1
AKs	.30%	330.5 : 1
2 Unpaired Big Cards	7.2%	12.8 : 1
Any Ace	14.50%	5.9 : 1
Rags	81%	.23 : 1

*Note – Suited connectors calculated are **45** through **TJ**. **2 Big Cards** refers to any 2 unpaired cards **J** or higher.

The Calculations

These odds and probabilities can be calculated at least two ways:

1. Combinations — Knowing the total number of possible 2-card combinations (**1,326**) and the number

of combinations that will make the starting hand, the odds are easily calculated as demonstrated below.

2. Probabilities — The probability of any starting hand can easily be determined by multiplying the probability of the first card by the probability of the second.

As demonstrated in Chapter 1 — *Basic Calculations* — and throughout this book, any odds are easily converted into probabilities and *vice versa*.

Pocket Pairs

For each rank — cards **2** through **A**, there are 6 possible pair combinations — **Comb(4, 2)**:

$$\frac{(4 * 3)}{(1 * 2)} = 6$$

With **13** ranks of **6** possible pair combinations there are **78 pair combinations** possible from a standard 52-card deck.

The odds and probability that any given player will receive the Pairs listed below as his hole cards are:

Hand	Probability	Odds
Any Pair	5.9%	16 : 1
Any Specific Pair (AA)	.452%	220 : 1
Pair of Jacks or Better	1.8%	54.25 : 1
22 through TT	4.10%	23.6 : 1

Pocket Aces

There are 6 different 2-card combinations that **WILL** contain a pair of Aces — **Comb(4, 2)**:

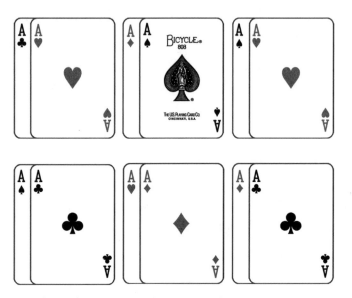

The number of 2-card combinations that **WILL NOT** contain a pair of Aces is **Total Possibilities (1,326) — WILLs** (6) = **1,320 WILLNOTs**.

The odds against being dealt Aces in the hole in Hold'em are:

WILLNOTs (1,320) : WILLs (6)
1,320 : 6
Reduce
1,320 / 6 : 6 / 6
220 : 1

Another way is to determine the probability of being dealt Aces in the hole, by multiplying the probability of being dealt the first Ace by the probability of being dealt the second Ace:

$$\frac{4}{52} * \frac{3}{51} = .0045249$$

This probability, expressed as a percentage (**.0045249 * 100**), is less than one half of one percent or **.45249%**.

As a check, we can convert the probability expressed as a decimal ratio to 1 (**.0045249**) back to odds:

$$\frac{\text{WILLs}}{\text{Total Possibilities}} = \text{Probability as decimal ratio to 1}$$

Odds = WILLNOTs : WILLs

Total Possibilities = 1
− WILLs = .0045249
WILLNOTs = .9954751

WILLNOTs : WILLs
.9954751 : .0045249
Reduce
.9954751 / .0045249 : .0045249 / .0045249
220 : 1

Any Pocket Pair

The likelihood of any Pair in the hole for any given player expressed as odds is:

Total Possibilities = 1,326

Total Possible Pairs (WILLs) = 78

Total Possibilities = 1,326
− WILLs = 78
WILLNOTs = 1,248

$$\text{WILLNOTs (1,248) : WILLs (78)}$$
$$1,248 : 78$$
$$\textbf{Reduce}$$
$$1,248 \text{ / } 78 : 78 \text{ / } 78$$
$$16 : 1$$

The likelihood of any Pair in the hole expressed as a probability is

$$\frac{\textbf{WILLs}}{\textbf{Total Possibilities}} * 100 = \textbf{Probability as \%}$$

$$\frac{78}{1,326} * 100 = 5.9\%$$

On average you will be dealt a Pocket Pair once in **17** hands or roughly **6%** of the time.

Pocket JJ or Better

For **JJ** through **AA** there are **24** (**6 * 4**) possible pair combinations from **1,326** total possible 2-card combinations:

$$\textbf{Possible Pairs JJ or Better (WILLs) = 24}$$

$$\textbf{Total Possibilities = 1,326}$$
$$\textbf{– WILLs = 24}$$
$$\textbf{WILLNOTs = 1,302}$$

$$\textbf{WILLNOTs (1,302) : WILLs (24)}$$
$$1,302 : 24$$
$$\textbf{Reduce}$$
$$1,302 \text{ / } 24 : 24 \text{ / } 24$$
$$54.25 : 1$$

The probability of a pair of Jacks or better in the hole is

$$\frac{\text{WILLs}}{\text{Total Possibilities}} * 100 = \text{Probability as \%}$$

$$\frac{24}{1,326} * 100 = 1.8\%$$

On average you will be dealt Pocket Jacks or better once in **55.25** hands or roughly **2%** of the time.

Pocket 22 through TT

For **22** through **TT** there are **54** (**9 * 6**) possible pair combinations from **1,326** total possible 2-card combinations.

Possible Pairs 22 through TT (WILLs) = 54

Total Possibilities = 1,326
– WILLs = 54
WILLNOTs = 1,272

WILLNOTs (1,272) : WILLs (54)
1,272 : 54
Reduce
1,272 / 54 : 54 / 54
23.6 : 1

The likelihood of pocket **22** through **TT**, expressed as a probability, is shown below — first as a percentage calculation then as a combination of probabilities.

Percentage calculation:

$$\frac{\text{WILLs}}{\text{Total Possibilities}} * 100 = \text{Probability as \%}$$

$$\frac{54}{1,326} * 100 = 4.1\%$$

Combination of Probabilities:

There are **36** cards in the deck with a rank of **2** through **T**, but there are only 4 cards of each rank from which your Pair can be dealt. The probability of the first card is **36/52** and the probability of the second card is **3/51**:

$$\frac{36}{52} * \frac{3}{51} = .0407$$

It is very easy to calculate odds and probabilities using these methods.

On average you will be dealt pocket **22** through **TT** once in **24.6** hands or roughly **4.1%** of the time.

Suited Cards

Given enough action, suited cards — especially suited connectors — can make great drawing hands.

The table below reflects the likelihood that any given player will hold any two suited cards, suited connectors or a suited Ace (**Ax**).

Hand	Probability	Odds
Suited Connectors	2.10%	46.4 : 1
Suited Ace	3.62%	26.6 : 1
Any 2 Suited	23.53%	3.25 : 1
*Note — Suited connectors calculated are **45** through **TJ**.		

Suited Connectors

These are the connectors that will make a Straight when combined with 3 cards in either or both directions.

There are **7** ranks and **4** suits so **7 * 4 = 28** possible suited connectors **45** through **TJ** from **1,326** total possible 2-card combinations:

Suited Connectors 45 through TJ (WILLs) = 28

Total Possibilities = 1,326
– WILLs = 28
WILLNOTs = 1,298

WILLNOTs (1,298) : WILLs (28)
1,298 : 28
Reduce
1,298 / 28 : 28 / 28
46.4 : 1

The probability of suited **45** through **TJ** in the hole is

$$\frac{\textbf{WILLs}}{\textbf{Total Possibilities}} * 100 = \textbf{Probability as \%}$$

$$\frac{28}{1,326} * 100 = 2.1\%$$

On average you will be dealt one of these sets of suited connectors once in **47.4** hands or roughly **2.1%** of the time.

Any Suited Ace – Axs

There are **12 Ax** combinations for each of the 4 suits, resulting in **48** possible **Ax** suited combinations:

Axs Combinations (WILLs) = 48

Total Possibilities = 1,326
– WILLs = 48
WILLNOTs = 1,278

WILLNOTs (1,278) : WILLs (48)
1,278 : 48
Reduce
1,278 / 48 : 48 / 48
26.6 : 1

The probability of **Axs** in the hole is

$$\frac{\textbf{WILLs}}{\textbf{Total Possibilities}} * 100 = \textbf{Probability as \%}$$

$$\frac{48}{1,326} * 100 = 3.6\%$$

On average you will be dealt **Axs** once in **27.6** hands or roughly **3.6%** of the time.

Any 2 Suited

Because the first card can be any card in the deck, the probability that you will be dealt 2 suited cards is equal to the probability that the suit of the second card will match the suit of the first:

$$\frac{\textbf{WILLs}}{\textbf{Total Possibilities}} * 100 = \textbf{Probability as \%}$$

$$\frac{12}{51} * 100 = 2.1\%$$

To convert this probability to odds:

Total Possibilities = 51
– WILLs = 12
WILLNOTs = 39

WILLNOTs (39) : WILLs (12)
39 : 12
Reduce
39 / 12 : 12 / 12
3.25 : 1

To verify this somewhat different approach, calculate the same odds using sets and combinations.

To find the odds against 2 suited cards in the hole, first find the number of possible 2-card combinations for each suit — **Comb(13, 2)**, then multiply by **4**:

$$\frac{(13 * 12)}{(1 * 2)} * 4 = 312$$

WILLs = 312

Total Possibilities = 1,326
– WILLs = 312
WILLNOTs = 1,014

WILLNOTs (1,014) : WILLs (312)
1,014 : 312
Reduce
1,014 / 312 : 312 / 312
3.25 : 1

Your hole cards will be suited once every **4.25** hands or on average **23.5%** of the time.

2 Big Cards

This table shows the odds that any given player's hole cards will be certain Big Card combinations.

Hand	Probability	Odds
Big Slick – Any AK	1.2%	81.9 : 1
Suited Big Slick – AKs	.3%	330.5 : 1
2 Unpaired Big Cards	7.2%	12.8 : 1
Any Ace	14.5%	5.9 : 1

Big Slick – Any AK

There are **4 * 4 = 16** 2-card combinations that WILL make any **AK** including **AKs**.

Total Possibilities = 1,326
– WILLs = 16
WILLNOTs = 1,310

WILLNOTs (1,310) : WILLs (16)
1,310 : 16
Reduce
1,310 / 16 : 16 / 16
81.9 : 1

The probability of Big Slick in the hole is:

$$\frac{\text{WILLs}}{\text{Total Possibilities}} * 100 = \text{Probability as \%}$$

$$\frac{16}{1,326} * 100 = 1.2\%$$

On average you will be dealt Big Slick once in **82.9** hands or roughly **1.2%** of the time. This includes the possibility that the hand will be suited.

Suited Big Slick – AKs

There are only **4** 2-card combinations that **WILL** make **AK** suited:

Total Possibilities = 1,326
– WILLs = 4
WILLNOTs = 1,322

WILLNOTs (1,322) : WILLs (4)
1,322 : 4
Reduce
1,322 / 4 : 4 / 4
330.5 : 1

The probability of suited Big Slick in the hole is

$$\frac{\text{WILLs}}{\text{Total Possibilities}} * 100 = \text{Probability as \%}$$

$$\frac{4}{1,326} * 100 = .3\%$$

On average you will be dealt Big Slick suited once in **331.5** hands or roughly **.3%** of the time.

2 Unpaired Big Cards

These are the number of possible 2-card combinations that contain 2 unpaired cards of Jack or better.

Among the **16** cards in the deck with a rank of Jack or better there are **120** possible combinations calculated as **Comb(16, 2)**:

$$\frac{(16 * 15)}{(1 * 2)} = 120$$

This number includes **24** possible pair combinations. There are **120 – 24 = 96** combinations that **WILL** contain 2 unpaired cards above the rank of Ten.

Another way is to multiply the probability of the first card by the probability of the second. In this case, after the first card is dealt, there are only **12** cards left that will not pair the first card:

$$(16 * 12) / (1 * 2) = 96 \text{ (WILLs)}$$

Total Possibilities = 1,326
– WILLs = 96
WILLNOTs = 1,230

WILLNOTs (1,230) : WILLs (96)
1,230 : 96
Reduce
1,230 / 96 : 96 / 96
12.8 : 1

The probability of 2 unpaired Big Cards in the hole is

$$\frac{\textbf{WILLs}}{\textbf{Total Possibilities}} * 100 = \textbf{Probability as \%}$$

$$\frac{96}{1,326} * 100 = 7.2\%$$

On average you will be dealt 2 unpaired Big Cards once in **13.8** hands or roughly **7.2%** of the time.

Any Ace

There are **4 * 48 = 192** 2-card combinations that **WILL** make a hand that includes **Ax**:

Total Possibilities = 1,326
– WILLs = 192
WILLNOTs = 1,134

WILLNOTs (1,134) : WILLs (192)
1,134 : 192
Reduce
1,134 / 192 : 192 / 192
5.9 : 1

The probability of any Ace in the hole is

$$\frac{\textbf{WILLs}}{\textbf{Total Possibilities}} * 100 = \textbf{Probability as \%}$$

$$\frac{192}{1,326} * 100 = 14.5\%$$

On average you will be dealt an Ace once in every **6.9** hands or roughly **15%** of the time.

Rags

Rags are all the hands not mentioned above. Rags are the hands that are left over after all the good hands have been taken out. Rags are usually not playable except in special situations or for free from the blind.

With a given *caveat* that some small number of hands might be duplicated, i.e., suited connectors and 2 Big Cards, we will consider the number of rag hands to be the total number of possible hands (**1,326**) minus all of the non-rag hands.

Non-Rag Hands

The **WILLNOTs** are all of the non-rag hands:

> 78 – **Pair Hands**
> 28 – **Suited Connectors**
> 48 – **Suited Ace**
> 96 – **2 Unpaired Big Cards**
> 250 – **Non-Rag Hands (WILLNOTs)**

To find the likelihood of rags in the hole, subtract the total number of non-rag hands (**WILLNOTs**) from the **Total Possible** hands to find the number of possible rag hands (**WILLs**):

> **Total Possibilities = 1,326**
> **– WILLs = 250**
> **WILLNOTs = 1,076**
>
> **WILLNOTs (1,076) : WILLs (250)**
> **1,076 : 250**
> **Reduce**
> **1,076 / 250 : 250 / 250**
> **.23 : 1**

The probability of rags in the hole is

$$\frac{\textbf{WILLs}}{\textbf{Total Possibilities}} * \textbf{100} = \textbf{Probability as \%}$$

$$\frac{1,076}{1,326} * 100 = 81.2\%$$

On average you will be dealt rags **4** out of **5** hands or slightly over **81%** of the time.

Before the Flop

After the hole cards have been dealt and before the Flop some primary concerns are

1. Relative strength of the hand

2. Likelihood the hand will improve

3. Likely strength of competition

4. Estimated cost to continue in the hand

5. Estimated size of the final pot

6. Possible moves to eliminate competitors.

The section on starting hands gives a good idea of the relative rarity/strength of certain holdings and the likelihood that opponents will hold those same hands.

For example, from the discussion on starting hands we can see that the probability of being dealt a Pocket Pair is 1/17. The work below reveals that the probability of flopping a Set or better with a Pair in the hole is 1/8.51. By multiplying these probabilities it becomes known that only once in approximately every 145 hands you or any other specific player will flop a **hidden** Set.

With an average of 20–40 hands dealt per hour in casino poker and 50–70 hands dealt per hour in online poker, on average you will flop a Set or better to a Pocket Pair only once every 4–5 hours in the casino and once every 2–3 hours online.

Important to note from a game theory perspective: On average, someone at each table (but not necessarily you)

will flop this same hidden Set once every 12–18 minutes online and once every 25–40 minutes in the casino.

Most pre-Flop odds calculations relate to the likelihood that certain hands will improve. To help organize these calculations, almost all possible starting hands can be placed into one of four classifications:

1. Pocket Pairs

2. 2 Suited Cards

3. Big Cards

4. Rags.

Most of the work in this section refers to the odds of improvement of certain hands. Players generally find "odds of improvement" easier to work with than "probability of improvement" because, for them, it is **easier to relate those odds to the money odds in order to determine expectation.**

Any expression of odds can easily be converted to an expression of probability. For example, with a Pair in the pocket, the odds against flopping a Set and a Set only are **8.28 : 1.**

To convert these odds to an expression of probability:

WILLNOTs : WILLs
8.28 : 1

WILLs = 1
+ WILLNOTs = 8.28
Total Possibilities = 9.28

$$\frac{\textbf{WILLs}}{\textbf{Total Possibilities}} * 100 = \textbf{Probability as \%}$$

$$\frac{1}{9.28} * 100 = 10.8\%$$

The odds of **8.28 : 1** and the probability of **10.8%** are two different ways of saying the same thing.

This table reflects the odds of improvement on the Flop for some of the most common starting hands, followed by tables and text that demonstrate their calculation.

Starting Hand	Hand to Flop	Odds
Pocket Pair	Set Only	8.28 : 1
Pocket Pair	Set or Boat	7.5 : 1
Pocket Pair	Boat	101.1 : 1
Pocket Pair	Quads	407.3 : 1
Pocket Pair	Set or Better	7.3 : 1
Pocket Pair	2 Pair	5.19 : 1
Suited Connectors	Flush Draw	8.14 : 1
Suited Connectors	Flush	117.8 : 1
Suited Connectors	Straight Draw	9.2 : 1
Suited Connectors	Straight	75.56 : 1
Suited Connectors	Straight Flush Draw	141 : 1
Suited Connectors	Straight Flush	4,899 : 1
2 Suited	Pair Either	2.45 : 1
2 Suited	Pair Both	48.49 : 1
2 to Royal Flush	Royal Flush Draw	138 : 1
2 to Royal Flush	Royal Flush	19,599 : 1
Ax Suited	Pair of Aces	5 : 1
Ax Suited	Ace Flush Draw	8.14 : 1

Starting Hand	Hand to Flop	Odds
Ax Suited	Ace High Flush	117.8 : 1
2 Unpaired Cards	Pair a Certain Hole Card	5 : 1
2 Unpaired Cards	Pair Either Hole Card	2.45 : 1
2 Unpaired Cards	Pair Both Hole Cards	48.49 : 1
2 Unpaired Cards	A Set of Either Hole Cards	73.24 : 1

The Calculations

One number that will be used throughout these calculations is the total number of possible 3-card Flops. Using the basic formula for combinations and given your 2 hole cards, determine the total number of possible **3**-card Flops from the remaining **50** cards — **Comb(50, 3)**:

$$\frac{(50 * 49 * 48)}{(1 * 2 * 3)} = 19{,}600$$

Pocket Pairs

With a Pair in the pocket it is not possible to flop a Straight or a Flush. The possible improvements to a Pocket Pair on the Flop are two Pair, a Set, a Full House (Boat) or Quads.

Starting Hand	Hand to Flop	Odds
Pocket Pair	Set Only	8.28 : 1
Pocket Pair	Set or Boat	7.5 : 1

Starting Hand	Hand to Flop	Odds
Pocket Pair	**Boat**	**101.1 : 1**
Pocket Pair	**Quads**	**407.3 : 1**
Pocket Pair	**Set or Better**	**7.3 : 1**
Pocket Pair	**2 Pair**	**5.19 : 1**

Pocket Pair >>> Set Only

To find the odds of flopping a Set and a Set only we need to find the number of possible Flops that will include one of the cards to make the Set and no Pair.

First find the number of 2-card combinations possible from the remaining unseen **50** cards minus the **2** cards that will make the Set — **Comb(48, 2)**:

$$\frac{(48 * 47)}{(1 * 2)} = 1,128$$

There are 1,128 2-card combinations that do not include the 2 cards either one of which will make the Set. From this number subtract all of the possible 2-card combinations in the deck that are Pairs (78) minus the 6 Pairs that come from the rank of the hole cards (**72**):

$$1,128 - 72 = 1,056$$

Multiply this number (**1,056**) by the number of remaining cards that **WILL** make the Set (**2**) and you will have the number of possible 3-card Flops that will make the Set and the Set only:

$$1,056 * 2 = 2,112$$

We now have all the numbers needed to calculate the odds of flopping a Set and Set only:

Total Possibilities = 19,600
– WILLs = 2,112
WILLNOTs = 17,488

Odds of Flopping a Set Only with a Pocket Pair
WILLNOTs : WILLs
17,488 : 2,112
Reduce
17,488 / 2,112 : 2,112 / 2,112
8.28 : 1

Pocket Pair >>> Set or Full House

This calculation is much the same as the one above except you don't subtract the possible Pairs from the possible 2-card combinations.

First find the number of 2-card combinations possible from the remaining unseen **50** cards minus the **2** cards that will make the Set — **Comb(48, 2)**:

$$\frac{(48 * 47)}{(1 * 2)} = 1,128$$

Multiply the result (**1,128**) by the number of remaining cards that **WILL** make the Set (**2**) and you will have the number of possible 3-card Flops that will make the Set or a Full House:

$$1,128 * 2 = 2,256$$

Also there is the possibility that a Set could come on the Flop, which would also make a Full House with the Pair in the pocket. From the 12 ranks that do not include the rank of the Pocket Pair there are **48** different 3-card combinations that are Sets — **Comb(4, 3)**:

$$\frac{(4 * 3 * 2)}{(1 * 2 * 3)} * 12 = 48$$

In all there are **2,256 + 48 = 2,304** possible 3-card Flops that will make a Set or Full House when holding a Pocket Pair.

We now have all the numbers needed to calculate the odds of flopping a Set or a Full House:

Total Possibilities = 19,600
– WILLs = 2,304
WILLNOTs = 17,296

Odds of Flopping a Set or Boat with a Pocket Pair
WILLNOTs : WILLs
17,296 : 2,304
Reduce
17,296 / 2,304 : 2,304 / 2,304
7.5 : 1

Pocket Pair >>> Full House

To flop a Full House with a Pair in the hole, the 3-card board must contain 1 of the remaining 2 cards that make the Set and any of the remaining **72** Pairs; or the Flop itself must be a Set. Any of the **72** Pairs may combine with either of the **2** cards that will make the Set to make the Full House, which means that there are:

72 * 2 = 144

plus the **48** possible Sets that can come on the Flop to make **144 + 48 = 192** possible 3-card Flops that will make a Full House when holding a Pocket Pair.

We now have all the numbers needed to calculate the odds of flopping a Full House with a Pair in the hole:

Total Possibilities = 19,600
– WILLs = 192
WILLNOTs = 19,408

Odds of Flopping a Boat with a Pocket Pair
WILLNOTs : WILLs
19,408 : 192
Reduce
19,408 / 192 : 192 / 192
101.1 : 1

Pocket Pair >>> Quads

To flop Quads with a Pair in the hole, 2 of the three Flop cards must be of the same rank as the Pocket Pair and the third card can be any of the remaining **48** unseen cards. With only **1** 2-card combination for the first two cards and **48** possibilities for the third card, the number of possible 3-card Flops that **WILL** make Quads is

1 * 48 = 48

We now have all the numbers needed to calculate the odds of flopping 4 of a kind with a Pair in the hole:

Total Possibilities = 19,600
– WILLs = 48
WILLNOTs = 19,552

Odds of Flopping Quads with a Pocket Pair
WILLNOTs : WILLs
19,552 : 48
Reduce
19,552 / 48 : 48 / 48
407.3 : 1

Pocket Pair >>> Set or Better

Given all of the above calculations, finding the odds of flopping a Set, a Full House or Quads when holding a Pocket Pair is simply a matter of combining the **WILLs**.

Number of Flops that make a Set	**2,112**
Number of Flops that make a Boat	**192**
Number of Flops that make Quads	**48**
Flops that WILL make a Set or better	**2,352**

We now have all the numbers needed to calculate the odds of flopping a Set or better with a Pair in the hole:

Total Possibilities = 19,600
− WILLs = 2,352
WILLNOTs = 17,248

Odds of Flopping Set or Boat w/Pocket Pair
WILLNOTs : WILLs
17,248 : 2,352
Reduce
17,248 / 2,352 : 2,352 / 2,352
7.3 : 1

Pocket Pair >>> 2 Pair

To flop 2 Pair with a Pair in the hole, the Flop must contain a Pair and a third card that doesn't match the Pocket Pair or the other cards on the board.

Without the rank of the Pocket Pair, there are **72** Pairs among the remaining **50** unseen cards.

To find the number of Flops that will make 2 Pair with a Pair in the pocket, multiply the number of possible Pairs (**72**) by the number of remaining unseen cards that will not match either the Pair in the hole or the Pair on the board, and that will not be the pair on the board (**50 − 2 − 2 − 2 = 44**):

$$72 * 44 = 3,168$$

We now have all the numbers needed to calculate the odds of flopping 2 Pair with a Pair in the hole:

Total Possibilities = 19,600
− WILLs = 3,168
WILLNOTs = 16,432

Odds of Flopping 2 Pair w/Pocket Pair
WILLNOTs : WILLs
16,432 : 3,168
Reduce
16,432 / 3,168 : 3,168 / 3,168
5.19 : 1

Suited Cards

While any 2 suited cards can make a Flush, the most valuable suited cards are the suited connectors, and the higher the rank, the better. For our purposes, we will consider the suited connectors that range from **45** through **TJ**. These are the connectors that can make a Straight in both directions.

Ace–rag suited is a hand played far too often by beginning players. As we will demonstrate, this hand usually will not flop a Flush draw and when it does, most of the time it will not make the Flush. When the player does flop an Ace, he is left trapped with a top Pair and a weak kicker — a hand that can be a very expensive second-best.

The calculation of the odds that you will pair either or both of your two suited cards are contained in the section below on "*2 Big Cards*."

2 Suited in the Hole

Starting Hand	Hand to Flop	Odds
Suited Connectors	Flush Draw	8.14 : 1
Suited Connectors	Flush	117.8 : 1
Suited Connectors	Straight Draw	9.2 : 1
Suited Connectors	Straight	75.56 : 1
Suited Connectors	Straight Flush Draw	141 : 1
Suited Connectors	Straight Flush	4,899 : 1
2 Suited	Pair Either	2.45 : 1
2 Suited	Pair Both	48.49 : 1
2 to Royal Flush	Royal Flush Draw	138 : 1
2 to Royal Flush	Royal Flush	19,599 : 1
Ax Suited	Pair of Aces	5 : 1
Ax Suited	Ace Flush Draw	8.14 : 1
Ax Suited	Ace High Flush	117.8 : 1

Suited Connectors

Suited connectors are hands that can offer great opportunity. A Rags Flop with small to medium suited connectors can sometimes leave the player in a dream situation: a big hand or big draw that is completely disguised.

The primary *caveat* for playing these hands, especially the connectors of medium to low rank, is that the player must have either

♦ a firm indication that there will be a large enough final pot to make the draw worthwhile, or

♦ a strong indication that some or all of the remaining players might fold to a properly executed bluff or semi-bluff.

2 Suited >>> Flush Draw

With 2 suited cards in the hole, calculate the odds of flopping a Flush draw by first finding the number of suited 2-card combinations possible from the remaining **11** suited cards — **Comb(11, 2)**, and multiplying it by the number of unseen cards not of that suit (**52 − 13 = 39**), resulting in the number of Flops that **WILL** make a Flush draw.

Next, subtract the **WILLs** from the total number of possible Flops (**19,600**) to find the **WILLNOTs**. Reduce the result and you have the odds of flopping a Flush draw with 2 suited cards in the hole:

$$\frac{(11 * 10)}{(1 * 2)} = 55$$

$$39 * 55 = 2,145$$

We now have all the numbers needed to calculate the odds of flopping a Flush draw with 2 suited cards in the hole:

Total Possibilities = 19,600
– WILLs = 2,145
WILLNOTs = 17,455

Odds of Flopping Flush Draw with 2 Suited
WILLNOTs : WILLs
17,455 : 2,145
Reduce
17,455 / 2,145 : 2,145 / 2,145
8.14 : 1

2 Suited >>> Flush

To calculate the odds against flopping a Flush with 2 suited cards in the hole, first find the number of possible 3-card Flops where all three cards match the suit of the hole cards — **Comb(11, 3)**:

$$\frac{(11 * 10 * 9)}{(1 * 2 * 3)} = 165$$

We now have all the numbers needed to calculate the odds of flopping a Flush with 2 suited cards in the hole:

Total Possibilities = 19,600
– WILLs = 165
WILLNOTs = 19,435

Odds of Flopping Flush with 2 Suited
WILLNOTs : WILLs
19,435 : 165

Reduce
19,435 / 165 : 165 / 165
117.8 : 1

Suited Connectors >>> Open Ended Straight Draw

For this calculation we will consider suited connectors from **45** through **TJ**.

With **7♦8♦** in the hole there are **3** 2-card combinations that can appear on the Flop to make a Straight draw:

From each of these 2-card combinations there are **16** possible combinations. Therefore, there are **48** possible 2-card combinations that can be part of the Flop that will make an open ended Straight draw. This also includes the possibility that the Flop will pair either a hole card or a board card and that the open ended Straight draw might be a Straight Flush draw.

After you subtract the **2** hole cards, the **2** cards that make the draw, and the **8** cards that would complete the Straight, there are **40** unseen cards left.

With the possibility that the Flop will pair either a hole card or a board card, the number of possible 3-card Flops that **WILL** make an open ended Straight draw to one of the suited connectors is

$$40 * 48 = 1{,}920$$

We now have all the numbers needed to calculate the odds of flopping an open ended Straight draw with 2 connected cards in the hole:

Total Possibilities = 19,600
– WILLs = 1,920
WILLNOTs = 17,680

Odds of Flopping Open Ended Straight Draw
WILLNOTs : WILLs
17,680 : 1,920
Reduce
17,680 / 1,920 : 1,920 / 1,920
9.2 : 1

Suited Connectors >>> Straight

With one of the above-mentioned suited connectors in the hole, to calculate the odds of flopping a Straight, including a Straight Flush, first find the number of possible 3-card combinations that will flop a Straight. With **7♦8♦** in the hole, there are **4** possible 3-card combinations that will make a Straight:

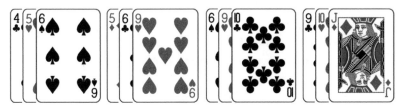

For each of these there are **4 * 4 * 4 = 64** possible combinations. Thus the number of possible 3-card Flops that will make the Straight is:

$$4 * 64 = 256$$

We now have all the numbers needed to calculate the odds of flopping a Straight with 2 connected cards in the hole:

Total Possibilities = 19,600
– WILLs = 256
WILLNOTs = 19,344

Odds of Flopping a Straight
WILLNOTs : WILLs
19,344 : 256
Reduce
19,344 / 256 : 256 / 256
75.56 : 1

Suited Connectors >>> Open Ended Straight Flush Draw

With one of the above-mentioned suited connectors in the hole, there are only **3** 2-card combinations that will flop an open ended Straight Flush draw.

With **7♥8♥** in the hole there are **3** 2-card combinations that can appear on the Flop to make a Straight Flush draw:

Minus the **2** hole cards and the **2** cards that make the draw and minus the **2** cards that would complete the Straight Flush, there are **46** unseen cards.

With the possibility that the Flop will pair either a hole card or a board card, the number of possible 3-card Flops that **WILL** make an open ended Straight Flush draw to one of the above-mentioned suited connectors is:

$$3 * 46 = 138$$

We now have all the numbers needed to calculate the odds of flopping an open ended Straight Flush draw with 2 suited and connected cards in the hole:

Total Possibilities = 19,600
– WILLs = 138
WILLNOTs = 19,462

Odds of Flopping Open Ended Straight Draw
WILLNOTs : WILLs
19,462 : 138
Reduce
19,462 / 138 : 138 / 138
141 : 1

Suited Connectors >>> Straight Flush

To find the odds of flopping a Straight Flush with one of the above-mentioned suited connectors in the hole, first find the number of possible 3-card combinations that will flop a Straight Flush.

With **7♥8♥** in the hole there are **4** possible 3-card combinations that will make a Straight Flush:

We now have all the numbers needed to calculate the odds of flopping a Straight Flush with 2 suited connected cards in the hole:

Total Possibilities = 19,600
– WILLs = 4
WILLNOTs = 19,596

Odds of Flopping a Straight Flush
WILLNOTs : WILLs
19,596 : 4
Reduce
19,596 / 4 : 4 / 4
4,899 : 1

2 Royal Cards >>> Royal Flush Draw

With 2 cards to a Royal in the hole there are only **3** 2-card combinations that can appear on the Flop that will make a Royal Flush draw:

For example, with **J♥Q♥** in the hole these are the only **3** 2-card combinations that can come on the Flop to yield a Royal Flush draw:

Minus the **2** hole cards and the **2** cards that make the draw, and minus the **1** card that would complete the Royal Flush, there are **47** unseen cards.

With the possibility that the Flop will pair either a hole card or a board card, the number of possible 3-card Flops that **WILL** make a Royal Flush draw is:

$$3 * 47 = 141$$

We now have all the numbers needed to calculate the odds of flopping a Royal Flush draw with 2 Royal cards in the hole:

Total Possibilities = 19,600
– WILLs = 141
WILLNOTs = 19,459

Odds of Flopping a Royal Flush Draw
WILLNOTs : WILLs
19,459 : 141
Reduce
19,459 / 141 : 141 / 141
138 : 1

2 Royal Cards >>> Royal Flush

With 2 cards to a Royal Flush in the hole, there is only **1** 3-card flop that **WILL** make the Royal.

We now have all the numbers needed to calculate the odds of flopping a Royal Flush with 2 Royal cards in the hole:

Total Possibilities = 19,600
– WILLs = 1
WILLNOTs = 19,599

Odds of Flopping a Royal Flush
WILLNOTs : WILLs
19,599 : 1
Reduce
19,599 / 1 : 1 / 1
19,599 : 1

2 Big Cards

Big Cards are any 2 cards Jack or higher. If the cards are paired then their odds of improvement are noted above. The same holds true if the 2 cards are either connected or suited.

Below are the calculations that the Flop will pair one particular card in the hole, will pair either card in the hole, will pair both cards in the hole, or will make a Set to either hole card.

Starting Hand	Hand to Flop	Odds
2 Unpaired Cards	Pair a Certain Hole Card	5 : 1
2 Unpaired Cards	Pair Either Hole Card	2.45 : 1
2 Unpaired Cards	Pair Both Hole Cards	48.49 : 1
2 Unpaired Cards	A Set of Either Hole Cards	73.24 : 1

2 Unpaired Hole Cards >>> Pair a Specific Hole Card

To find the odds of flopping at least a pair of Aces with **AJ** in the hole, find the number of 2-card combinations that do not contain an Ace, and multiply that number by the number of remaining unseen Aces, resulting in the total number of possible 3-card combinations that will contain at least an Ace.

Subtract the **2** cards in the hole and the **3** remaining Aces, leaving **47** unseen cards none of which are Aces, then use **Comb(47, 2)**:

$$\frac{(47 * 46)}{(1 * 2)} = 1{,}081$$

With **1,081** possible 2-card combinations that contain no Aces and **3** remaining Aces unseen, the number of possible 3-card Flops that **WILL** flop at least a pair of Aces is

$$3 * 1{,}081 = 3{,}243$$

We now have all the numbers needed to calculate the odds of flopping at least a pair of Aces with **AJ** in the hole:

Total Possibilities = 19,600
– WILLs = 3,243
WILLNOTs = 16,357

Odds of Flopping at Least AA
WILLNOTs : WILLs
16,357 : 3,243
Reduce
16,357 / 3,243 : 3,243 / 3,243
5 : 1

2 Unpaired Hole Cards >>> Pair Either Hole Card

To find the odds of flopping at least a Pair of either hole card with **AJ** in the hole, find the number of 2-card combinations that do not contain either an Ace or a Jack, multiply that number by the number of remaining unseen Aces and Jacks, giving you the total number of possible 3-card combinations that will contain at least one Ace or one Jack.

Subtract the **2** cards in the hole and the 6 remaining Aces and Jacks, leaving **44** unseen cards none of which are Aces or Jacks, then use Comb**(44, 2)**:

$$\frac{(44 * 43)}{(1 * 2)} = 946$$

With **946** possible 2-card combinations that contain no Aces or Jacks plus **6** remaining Aces and Jacks, the number of possible 3-card Flops that **WILL** flop at least a pair of Aces or a pair of Jacks is

$$6 * 946 = 5,676$$

We now have all the numbers needed to calculate the odds of flopping at least a Pair of either Aces or Jacks with **AJ** in the hole:

Total Possibilities = 19,600
– WILLs = 5,676
WILLNOTs = 13,924

Odds of Flopping AA or JJ
WILLNOTs : WILLs
13,924 : 5,676
Reduce
13,924 / 5,676 : 5,676 / 5,676
2.45 : 1

2 Unpaired Hole Cards >>> Pair Both Hole Cards on the Flop

To find the odds of a Flop that will pair both hole cards with **AJ** in the hole, we know that 2 of the 3 Flop cards must be **AJ**. Find the number of possible **AJ** 2-card combinations from the remaining **3** Jacks and **3** Aces:

$$3 * 3 = 9$$

Minus the **2** cards in the hole and the **6** remaining Aces and Jacks there are **44** unseen cards, none of which are Aces or Jacks.

With **9** possible 2-card combinations that contain both an Ace and a Jack and **44** unseen cards that are neither an Ace or a Jack, the number of possible 3-card Flops that **WILL** contain at least one Ace and one Jack is

$$9 * 44 = 396$$

We now have all the numbers needed to calculate the odds of flopping both a pair of Aces and a pair of Jacks with **AJ** in the hole:

Total Possibilities = 19,600
– WILLs = 396
WILLNOTs = 19,204

Odds of a Flop That Pairs Both Hole Cards
WILLNOTs : WILLs
19,204 : 396
Reduce
19,204 / 396 : 396 / 396
48.49 : 1

2 Unpaired Hole Cards >>> Flop a Set

To find the odds of a Flop that will include either a pair of Aces or a pair of Jacks with **AJ** in the hole, we need to know the number of possible Pairs of both Aces and Jacks from the remaining **3** Aces and **3** Jacks:

$$\frac{(3 * 2)}{(1 * 2)} = 3$$

With **3** possible pairs of Aces and **3** possible pairs of Jacks there are **6** different Pairs that can appear as part of the Flop to make a Set to either hole card.

Minus the **2** cards in the hole and the **6** remaining Aces and Jacks there are **44** unseen cards, none of which are Aces or Jacks.

With **6** possible Pairs and **44** unseen cards that are not either an Ace or a Jack, the number of possible 3-card Flops that **WILL** flop a not-too-hidden Set of either hole card is:

$$6 * 44 = 264$$

We now have all the numbers needed to calculate the odds of flopping a Set to either hole card:

Total Possibilities = 19,600
– WILLs = 264
WILLNOTs = 19,336

Odds of Flopping a Set
WILLNOTs : WILLs
19,336 : 264
Reduce
19,336 / 264 : 264 / 264
73.24 : 1

After the Flop

Odds of Improvement

Once the Flop has hit the board, you have 5 of the 7 cards that will make your best and final hand. After the Flop and with 2 cards to come, regardless of the hand, any player still involved will want to know:

- ♦ The odds of improvement on the next card (the Turn)

- ♦ The odds of improvement over the next two cards (the Turn and the River)

- ♦ The odds of hitting certain Runner-Runner hands.

The tables and calculations below demonstrate these odds.

Money Odds, Overlays & Expectation

After the betting on the Flop in limit games, the stakes usually double and the cost to continue in the hand increases. As each card is placed on the board, the likelihood of improvement decreases. As each bet or raise is made or called, the money odds get better.

The odds of improvement make it easy for any player to know his chances for a stronger hand. But to be most useful as decision support information, the odds of improvement must be considered in conjunction with the

- ♦ Money odds and the risk odds offered if a bluff or semi-bluff were to immediately win the pot

- ♦ Likely total money odds at the end

♦ Likely total cost to continue to the end

♦ Total likelihood of winning the pot.

Like the casinos in Las Vegas, successful poker players make their money by operating only when the environment offers positive expectation.

If we consider the risk odds as the odds of whether or not you will win the pot, then when the money odds are greater than the risk odds, there is an overlay or, better said, a positive expectation.

Successful poker players make their money by being OVERPAID for the risks they take.

Sound conventional poker wisdom states: **A losing bet saved is the same as a bet won** with regard to its impact on a player's bottom line.

The same can be said of an opportunity of positive expectation passed.

With an 8:1 payout on a 4:1 risk there is an obvious positive expectation. While this proposition will certainly pay in the long run, the player will still lose money on 4 of 5 tries. As long as the overlay is present, to pass on such an opportunity has the same effect on the bottom line as making a series of bad plays. **Passing** on situations of significant positive expectation can be as harmful to the bottom line as **making** a series of plays that carry a significant negative expectation.

Just as the ability to recognize and make plays that carry a positive expectation is essential, another essential skill needed for long-term success is the ability to recognize and lay down a huge draw to a tiny reward — **an underlay**.

From a Game Theory Perspective

While it is important to know the money odds and the odds of improvement, to see the total picture a player must also consider certain aspects of opposing players' hands, as well as their likely actions and reactions.

To have a true idea of expectation in any given hand, a player needs to know only 3 pieces of information:

1. The likely size of the pot at the end

2. The cost to continue to the end

3. A player's **Total Odds** of winning the pot.

Sometimes calculating a player's **Total Odds** of winning the pot can be one of the most complex calculations in all of poker. This is because it involves far more than knowing the odds of whether or not a certain hand will improve. In addition, a player must also consider:

1. The possibility that he is drawing dead

2. The possibility that his hand is already the Nuts

3. The possibility that any of his opponents might either have him beat or out-draw him

4. The possibility that his opponents might be manipulated out of the hand by well placed bets or raises.

For example, assume you're in a game where you have good knowledge of your opponents, you are on the button before the last card and have a 25% or 1 in 4 chance of completing a draw to what will surely be the best hand.

A 25% probability of drawing the winning hand is equal to odds of **3 : 1** against. With odds of **3 to 1** against, a player must have money odds of **4 to 1** or better to justify a call.

In this same hand there are 2 opponents still involved in the pot. As a student of game theory, you've tracked these two opponents and calculated that there is a 50% chance they will both fold to a bet or raise.

In the above example, if a player only considers his odds of improvement, he has a 25% chance of winning and a 75% chance of losing and needs money odds of something considerably more than **3 to 1** to justify further participation in the hand.

On the other hand, if you consider your 25% chance of making the draw to win the pot and the 50% chance that your opponents will fold to a bet or raise, your **Total Odds** of winning the pot are approximately 60%. With a 60% chance of winning you can participate with money odds of **2 to 1** or even slightly less instead of the **4 to 1** required if you only consider your 25% probability of improving the hand.

While it's impossible to always accurately predict an opponent's hand or future actions, knowing basic odds and player history will give you a strong indication.

One of the great values of the application of game theory and the concept of **Total Odds** to poker is that it can help the astute player discover value plays with good positive expectation that might have otherwise remained hidden.

While it is harmful to your bottom line to play with an underlay, it is equally harmful to your long-term success to pass on value opportunities that offer significant positive expectation.

Operating with a thorough knowledge of your **Total Odds** of success and a good understanding of your opponents' likely actions or reactions, adds to both your long and short-term expectations.

Odds the Turn Card Will . . .

With certain 5-card holdings after the Flop and with their calculations demonstrated below, here are the odds that these holdings will improve with the laying of the Turn card:

After the Flop	Hand to Turn	Odds
Pair	Set	22.5 : 1
4 Suited	Flush	4.22 : 1
Open Straight Draw	Straight	4.9 : 1
Gut-shot St Draw	Straight	10.75 : 1
Open St & 4 Flush	Straight or Flush	2.1 : 1
1 Pr & Open St Draw	Set or Straight	3.7 : 1
1 Pr & 4 Flush	Set or Flush	3.3 : 1
2 Pair	Full House	10.75 : 1
Set	Full House or Better	5.7 : 1
2 Unpaired Cards	Pair Either Hole Card	6.8 : 1

With one card to come, the odds calculation for the various hands is quite straightforward and the only number that changes is the number of outs.

After the Flop and before the Turn, counting your two hole cards, there are **47** unseen cards.

Pair >>> Set

Whether the Pair is in the pocket or the player has paired one of his hole cards, there are still only **2** cards that will make the Set among the remaining 47. The odds of making a Set from a Pair on the Turn are:

Total Possibilities = 47
– WILLs = 2
WILLNOTs = 45

Odds of Turning a Set with a Pair
WILLNOTs : WILLs
45 : 2
Reduce
45 / 2 : 2 / 2
22.5 : 1

4 Flush >>> Flush

Again, whether there are 2 suited cards in the pocket or 1 suited card in the hole and 3 suited on the board, there are still **9** cards among the unseen 47 that will complete the Flush:

Total Possibilities = 47
– WILLs = 9
WILLNOTs = 38

Odds of Turning a Flush with a 4-Flush
WILLNOTs : WILLs
38 : 9
Reduce
38 / 9 : 9 / 9
4.22 : 1

Open Ended Straight Draw >>> Straight

Regardless of how the 4 cards making up the draw lie, there are **8** outs to make the Straight:

Total Possibilities = 47
– WILLs = 8
WILLNOTs = 39

Odds of Turning a Straight with an Open Ended Draw
WILLNOTs : WILLs
39 : 8
Reduce
39 / 8 : 8 / 8
4.9 : 1

Gut-shot Straight Draw >>> Straight

There are **4** cards among the remaining 47 that will complete the gut-shot draw:

Total Possibilities = 47
– WILLs = 4
WILLNOTs = 43

Odds of Turning a Straight with a Gut-shot Draw
WILLNOTs : WILLs
43 : 4
Reduce
43 / 4 : 4 / 4
10.75 : 1

Open Straight Draw & 4 Flush >>> Straight or Flush

With an open ended Straight draw there are **8** outs to the Straight. With a 4 Flush there are **9** cards that will make the Flush. Because **2** of these cards will make both the Straight and the Flush there are **9 + 8 − 2 = 15** outs to either the Straight or the Flush, if the straight draw is suited this calculation also includes the possibility of making a Straight Flush:

Total Possibilities = 47
− WILLs = 15
WILLNOTs = 32

Odds of Turning a Straight or a Flush
WILLNOTs : WILLs
32 : 15
Reduce
32 / 15 : 15 / 15
2.1 : 1

Pair & Open Straight Draw >>> Straight or Set

With a Pair and an open Straight draw there are **10** cards that will make either a Straight or a Set:

Total Possibilities = 47
− WILLs = 10
WILLNOTs = 37

Odds of Turning a Set or Straight
with a Pair and Open Straight Draw
WILLNOTs : WILLs
37 : 10

Reduce
37 / 10 : 10 / 10
3.7 : 1

Pair & 4 Flush >>> Flush or Set

With a Pair and a Flush draw there are **11** cards that will make either a Flush or a Set:

Total Possibilities = 47
– WILLs = 11
WILLNOTs = 36

Odds of Turning a Flush or Set
WILLNOTs : WILLs
36 : 11
Reduce
36 / 11 : 11 / 11
3.3 : 1

2 Pair >>> Full House

There are **4** cards among the remaining 47 that will make the Boat:

Total Possibilities = 47
– WILLs = 4
WILLNOTs = 43

Odds of Turning a Boat with 2 Pair
WILLNOTs : WILLs
43 : 4
Reduce
43 / 4 : 4 / 4
10.75 : 1

Set >>> Full House or Better

With a Set there is **1** card that will make Quads and **6** cards that will make a Boat:

Total Possibilities = 47
– WILLs = 7
WILLNOTs = 40

Odds of Turning a Boat or Better w/Set
WILLNOTs : WILLs
40 : 7
Reduce
40 / 7 : 7 / 7
5.7 : 1

Pair Either of 2 Unpaired Hole Cards

With 2 unpaired hole cards after the Flop there are **6** cards among the remaining unseen 47 cards that will make a Pair:

Total Possibilities = 47
– WILLs = 6
WILLNOTs = 41

Odds of Turning a Boat w/2 Unpaired Hole Cards
WILLNOTs : WILLs
41 : 6
Reduce
41 / 6 : 6 / 6
6.83 : 1

Odds with 2 Cards to Come

After the Flop and before the Turn card has been dealt there are 47 unseen cards. When calculating the odds with 2 cards to come, one important number is the total number of possible 2-card combinations that can be made from the remaining unseen 47 cards — **Comb(47, 2)**:

$$\frac{(47 * 46)}{(1 * 2)} = 1{,}081$$

With 2 cards to come there are 4 possibilities:

1. You will hit none of your outs on either card

2. You will hit one of your outs on the Turn

3. You will hit one of your outs on the River

4. You will hit one of your outs on the Turn and another on the River.

The most liberal of these calculations figures the odds of hitting an out on either the Turn or the River, or hitting an out on both the Turn and the River.

Odds of Hitting on Either or Both Turn & River

The odds of hitting at least one available out on either or both the Turn and the River are demonstrated below:

After the Flop	Hand After River Card	Odds
Pair	Set or Better	9.8 : 1
4 Suited	Flush	1.86 : 1
Open Straight Draw	Straight	2.18 : 1

After the Flop	Hand After River Card	Odds
Gut-shot St Draw	Straight	5.07 : 1
Open St & 4 Flush	Straight or Flush	.85 : 1
1 Pr & Open St Draw	Set or Straight	1.55 : 1
1 Pr & 4 Flush	Set or Flush	1.35 : 1
2 Pair	Full House or Better	5.1 : 1
Set	Full House or Better	1.95 : 1
2 Unpaired Cards	Pair or Better	3.1 : 1

To make these calculations, find the number of total 2-card combinations from the unseen 47 cards — **1,081**.

To find the **WILLs**, add the number of possible 2-card combinations that will contain 1 and **only 1** of the available outs to the number of 2-card combinations that will contain 2 out cards.

1 Pair >>> Set or Better

With a Pair, **2** of the unseen 47 cards will make a Set.

Thus, **2 * 45 = 90** is the number of 2-card combinations that will make a Set, including some that will also make a Pair, thus making a Full House.

There is **1** 2-card combination that is the Pair of the paired cards that will make Quads.

In addition, there are **9** possible pair combinations that will make a Full House to the Pair by making a Set with one of the 3 unpaired cards.

Of a possible 1,081 2-card combinations, **90 + 1 + 9 = 100 WILL** match with a Pair to make a Set or better:

Total Possibilities = 1,081
– WILLs = 100
WILLNOTs = 981

Odds of a Set or Better with 2 Cards to Come
WILLNOTs : WILLs
981 : 100
Reduce
981 / 100 : 100 / 100
9.8 : 1

4 Flush >>> Flush with 2 Cards to Come

Of 47 cards, **9** will make the Flush and **38** will not:

9 * 38 = 342

342 2-card combinations will contain 1 and only 1 Flush card.

$$\frac{(9 * 8)}{(1 * 2)} = 36$$

36 2-card combinations will contain 2 suited cards.

342 + 36 = 378 possible 2-card combinations **WILL** make the Flush:

Total Possibilities = 1,081
– WILLs = 378
WILLNOTs = 703

Odds of a Flush with 2 Cards to Come
WILLNOTs : WILLs
703 : 378
Reduce
703 / 378 : 378 / 378
1.86 : 1

Open Ended Straight Draw >>> Straight with 2 Cards to Come

Of 47 cards, **8** will make the Straight and **39** will not:

$$8 * 39 = 312$$

312 2-card combinations will contain 1 and only 1 Straight card.

$$\frac{(8 * 7)}{(1 * 2)} = 28$$

28 2-card combinations will contain 2 Straight cards including the possibility of a Pair.

312 + 28 = 340 possible 2-card combinations **WILL** make the Straight:

Total Possibilities = 1,081
– WILLs = 340
WILLNOTs = 741

Odds of a Straight with 2 Cards to Come
WILLNOTs : WILLs
741 : 340
Reduce

$$741 / 340 : 340 / 340$$
$$2.18 : 1$$

Gut-shot Straight Draw >>> Straight with 2 Cards to Come

Of 47 cards, **4** will make the Straight and **43** will not:

$$4 * 43 = 172$$

172 2-card combinations will contain 1 and only 1 Straight card.

$$\frac{(4 * 3)}{(1 * 2)} = 6$$

6 2-card combinations will contain 2 Straight cards that will put a Pair on the board.

172 + 6 = 178 possible 2-card combinations **WILL** make the Straight:

Total Possibilities = 1,081
– WILLs = 178
WILLNOTs = 903

Odds of Making a Gut-shot Straight Draw
with 2 Cards to Come
WILLNOTs : WILLs
903 : 178
Reduce
903 / 178 : 178 / 178
5.07 : 1

Open Straight Draw & 4 Flush >>> Straight or Flush with 2 Cards to Come

Of 47 cards, **8** will make the Straight and **9** will make the Flush. Of these **17** cards, 2 will make both the Straight and the Flush leaving **15** outs to either a Straight or a Flush. Any one of **15** cards will make either a Straight or a Flush and **32** will not:

$$15 * 32 = 480$$

480 2-card combinations will contain 1 and only 1 Straight card.

$$\frac{(15 * 14)}{(1 * 2)} = 105$$

105 2-card combinations will contain 2 Straight or Flush cards.

480 + 105 = 585 possible 2-card combinations that **WILL** make either or both the Straight and Flush:

Total Possibilities = 1,081
– WILLs = 585
WILLNOTs = 496

Odds of a Straight or Flush with 2 Cards to Come
WILLNOTs : WILLs
496 : 585
Reduce
496 / 585 : 585 / 585
.85 : 1

With an open-ended Straight draw and a Flush draw, most of the time the player will complete either/both the Straight or Flush draw.

These odds might be more easily understood and used when expressed as a probability.

To convert these odds to an expression of probability remember that odds are an expression of

WILLNOTs : WILLs

.85 : 1

and probability expressed as a percentage is:

$$\frac{\textbf{WILLs}}{\textbf{Total Possibilities}} \textbf{* 100 = Probability as \%}$$

WILLNOTs (.85) + WILL(1) = Total Possibilities (1.85)

$$\frac{\textbf{1}}{\textbf{1.85}} \textbf{* 100 = 54\%}$$

With the above draw, the player will make either/both a Straight or a Flush 54% of the time.

The odds of **.85 : 1** and the probability of **54%** are two different ways of saying the same thing.

Open Ended Straight Draw & Pair >>> Either or Both a Straight, a Set or Better with 2 Cards to Come

Of 47 cards, **8** will make the Straight, **2** will make the Set or better, and **37** will make neither:

10 * 37 = 370

370 2-card combinations will contain 1 and only 1 Straight card.

$$\frac{(10 * 9)}{(1 * 2)} = 45$$

45 2-card combinations will contain 2 Straight or Set cards including the possibility of a Boat or Quads.

In addition, there are **9** pair combinations that could make a Full House to the 3 unpaired cards.

370 + 45 + 9 = 424 possible 2-card combinations that **WILL** make the Straight, Set or Quads:

Total Possibilities = 1,081
− WILLs = 424
WILLNOTs = 657

Odds of a Straight, Set or Better with an Open Straight Draw & a Pair with 2 Cards to Come
WILLNOTs : WILLs
657 : 424
Reduce
657 / 424 : 424 / 424
1.55 : 1

4 Flush and a Pair >>> Either or Both a Flush, Set or Better with 2 Cards to Come

Of 47 cards, **9** will make the Flush, **2** will make the Set or better, and **36** will make neither:

$$11 * 36 = 396$$

396 2-card combinations will contain 1 and only 1 Flush card.

$$\frac{(11 * 10)}{(1 * 2)} = 55$$

55 2-card combinations will contain 2 Flush or Set cards including the possibility of a Boat or Quads.

In addition there are **9** pair combinations that could make a Full House to the 3 unpaired cards.

396 + 55 + 9 = 460 possible 2-card combinations that **WILL** make the Straight, Set, Boat or Quads:

Total Possibilities = 1,081
– WILLs = 460
WILLNOTs = 621

Odds of a Flush, Set, Boat or Quads
with a 4 Flush & a Pair with 2 Cards to Come
WILLNOTs : WILLs
621 : 460
Reduce
621 / 460 : 460 / 460
1.35 : 1

2 Pair >>> Full House or Better with 2 Cards to Come

Of 47 cards, **4** will make the Set or better and **43** will not:

4 * 43 = 172

172 2-card combinations will contain 1 card that will make a Full House.

There are 2 different Pairs or **2** 2-card combinations that will make Quads to 2 Pair.

In addition, there are **3** pair combinations that could make a Full House to the remaining single unpaired card.

172 + 2 + 3 = 177 possible 2-card combinations that **WILL** make a Set, Boat or Quads:

Total Possibilities = 1,081
- WILLs = 177
WILLNOTs = 904

Odds of a Set, Boat or Quads
with 2 Pair and 2 Cards to Come
WILLNOTs : WILLs
904 : 177
Reduce
904 / 177 : 177 / 177
5.1 : 1

Set >>> Full Boat or Better with 2 Cards to Come

Of 47 possible cards to come, **1** will make Quads to the Set and any of **6** will pair either of the 2 remaining cards to make a Full House. Thus any of **7** cards will improve the Set to a Boat or better and **40** will not:

$$7 * 40 = 280$$

280 possible 2-card combinations one of which **WILL** make a Boat or Quads.

$$\frac{(7 * 6)}{(1 * 2)} = 21$$

21 2-card combinations will contain 2 Boat or better cards.

Also among the unseen cards there are **66** possible Pairs that will make a Full House to the Set.

280 + 21 + 66 = 367 possible 2-card combinations that WILL make the Boat or Quads:

Total Possibilities = 1,081
− WILLs = 367
WILLNOTs = 714

Odds of a Boat or Quads
with a Set and 2 Cards to Come
WILLNOTs : WILLs
714 : 367
Reduce
714 / 367 : 367 / 367
1.95 : 1

Pair Either Unpaired Hole Card or Better with 2 Cards to Come

Of 47 possible cards to come, any of **6** will pair either of the 2 hole cards and **41** will not pair either:

6 * 41 = 246

246 possible 2-card combinations **WILL** make a Pair.

Also, there are **6** 2-card combinations (Pairs) that will make a Set to either hole card.

In addition, there are **3 * 3 = 9** 2-card combinations that will make 2 Pair.

246 + 6 + 9 = 261 possible 2-card combinations that **WILL** make a Pair or better:

Total Possibilities = 1,081
− WILLs = 261

WILLNOTs = 820

Odds to Pair Either Hole Card
WILLNOTs : WILLs
820 : 261
Reduce
820 / 261 : 261 / 261
3.14 : 1

Runner – Runner

After the Flop and with 2 cards to come, you will need help on both of them. **No one card will do it.**

Remember — runner-runners don't complete very often and in most cases require extraordinary money odds to be played profitably.

The table reflects the odds and the work below demonstrates the calculations:

After the Flop	With Runner – Runner	Odds
1 Pair	Full House or Better	37.6 : 1
1 Pair	Quads	1,080 : 1
2 Pair	Quads	539.5 : 1
RR Open Straight Draw	Straight	21.52 : 1
RR 1 Gap Straight Draw	Straight	32.8 : 1
RR 2 Gap Straight Draw	Straight	66.56 : 1

After the Flop	With Runner – Runner	Odds
Backdoor Flush	Flush	23 : 1
2 Unpaired Cards	Set to Either Hole Card	179.2 : 1
2 Unpaired Cards	Pair Both Hole Cards	119.5 : 1
Open RR Straight Flush Draw	Straight Flush	359.3 : 1

The Calculations

The 2 ways most often used to calculate the odds of hitting a runner-runner after the Flop are

1. Calculate the number of 2-card combinations that **WILL** do the job and compare that to the total possible 2-card combinations to find the number of 2-card combinations that **WILLNOT** complete the hand.

2. Multiply the probability of the first runner by the probability of the second runner and then convert that probability to odds.

Both approaches are demonstrated below.

For the combinations we need to know the **Total Possible** 2-card combinations from the remaining **47** unseen cards — **Comb(47, 2)**:

$$\frac{(47 * 46)}{(1 * 2)} = 1,081$$

1 Pair >>> Full House or Quads

There are **2** cards that will make a Set to the Pair, and **9** cards that will pair any of the 3 unpaired cards:

$$2 * 9 = 18$$

18 possible 2-card combinations **WILL** make a Set to the Pair and also a Pair to one of the 3 unpaired cards to make a Full House.

1 2-card combination will make Quads to the Pair.

9 2-card combinations (Pairs) will make a Set to any of the 3 unpaired cards to make the Full House.

18 + 1 + 9 = 28 possible 2-card combinations that **WILL** make a Boat or better:

Total Possibilities = 1,081
– WILLs = 28
WILLNOTs = 1,053

Odds of a RR Boat or Better with 1 Pair
WILLNOTs : WILLs
1,053 : 28
Reduce
1,053 / 28 : 28 / 28
37.6 : 1

1 Pair >>> Quads

There is only **1** 2-card combination that will make Quads to a Pair:

Total Possibilities = 1,081
– WILLs = 1
WILLNOTs = 1,080

Odds of RR Quads with a Pair
WILLNOTs : WILLs
1,080 : 1

2 Pair >>> Quads

There are only **2** 2-card combinations that will make Quads to either of 2 Pair:

Total Possibilities = 1,081
– WILLs = 2
WILLNOTs = 1,079

Odds of RR Quads with 2 Pair
WILLNOTs : WILLs
1,079 : 2
Reduce
1,079 / 2 : 2 / 2
539.5 : 1

Open Backdoor 3-Straight >>> Straight

With 3 connected cards above **2** and below **K** in rank, there are **3** sets of 2-card combinations that will make the Straight:

For each of these **3** sets there are **4 * 4 = 16** possible combinations. Therefore, there are a total of **3 * 16 = 48** 2-card combinations that will make the Straight shown:

Total Possibilities = 1,081
– WILLs = 48
WILLNOTs = 1,033

Odds of a RR Straight
WILLNOTs : WILLs
1,033 : 48
Reduce
1,033 / 48 : 48 / 48
21.52 : 1

1-Gap Backdoor 3-Straight >>> Straight

With a single-gap backdoor Straight draw above **2** and below **K** in rank, there are **2** sets of 2-card combinations that will make the Straight:

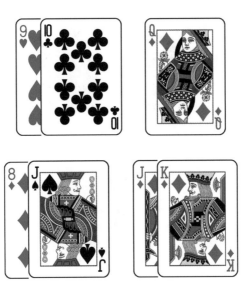

For each of the **2** sets there are **4 * 4 = 16** possible combinations. Thus, there are a total of **2 * 16 = 32** 2-card combinations that will make the Straight shown:

Total Possibilities = 1,081
– WILLs = 32
WILLNOTs = 1,049

Odds of a RR Straight
WILLNOTs : WILLs
1,049 : 32
Reduce
1,049 / 32 : 32 / 32
32.8 : 1

2-Gap Backdoor 3-Straight >>> Straight

With a 2-gap backdoor Straight draw above **2** and below **K** in rank, there is only **1** Set of 2-card combinations that will make the Straight:

With only **1** Set of 2-card combinations that will make the Straight, there are only **4 * 4 = 16** 2-card combinations that will make the Straight shown:

Total Possibilities = 1,081
– WILLs = 16
WILLNOTs = 1,065

Odds of a RR Straight
WILLNOTs : WILLs
1,065 : 16
Reduce
1,065 / 16 : 16 / 16
66.56 : 1

Backdoor Flush >>> Flush

With a 3-card Flush draw there are **10** cards of the right suit that remain among the 47 unseen cards. From those **10** suited cards there are **Comb(10, 2)**:

$$\frac{(10 * 9)}{(1 * 2)} = 45$$

45 possible 2-card combinations that **WILL** make a runner-runner Flush:

Total Possibilities = 1,081
− WILLs = 45
WILLNOTs = 1,036

Odds of a RR Flush
WILLNOTs : WILLs
1,036 : 45
Reduce
1,036 / 45 : 45 / 45
23 : 1

Either Unpaired Hole Card >>> Runner-Runner Set

There are **3** 2-card/pair combinations for each hole card that will make a Set. Thus there are **6** 2-card combinations that will make a Set to either unpaired hole card:

Total Possibilities = 1,081
− WILLs = 6
WILLNOTs = 1,075

Odds of RR Set to Either Unpaired Hole Card
WILLNOTs : WILLs
1,075 : 6
Reduce
1,075 / 6 : 6 / 6
179.2 : 1

2 Unpaired Hole Cards >>> Runner-Runner 2 Pair

With 2 unpaired cards in the hole there are **6** cards among the remaining 47 that will pair either card.

For this calculation we will multiply the probability that the first card will pair either card by the probability that the second card will pair the remaining unpaired card.

That probability expressed as a decimal ratio to 1 is:

$$\frac{\text{WILLs}}{\text{Total Possibilities}} = \text{Probability as decimal ratio to 1}$$

With **6** cards of a total of 47 that will pair either hole card, the probability that the first card will pair either hole card is:

$$\frac{6}{47}$$

With the first card paired there are **3** of the remaining 46 cards that will pair the remaining unpaired hole card. The probability that the second card will make the second Pair is

$$\frac{3}{46}$$

The probability of runner-runner pairing both unpaired hole cards is

$$\frac{6}{47} * \frac{3}{46} = \frac{18}{2,162}$$

$$\frac{18}{2,162} = .0083$$

The probability of runner-runner pairing both hole cards is **.0083** or **.83%** — less than **1%**.

To convert this decimal ratio to odds, subtract the **WILLs** (**.0083**) from the total possibility (**1.0**) to find the **WILLNOTs** = .9917.

By approximation we know that if the probability is just under **1%**, then the odds should be just over **100** to **1**:

Total Possibilities = 1
− WILLs = .0083
WILLNOTs = .9917

Odds of RR to Pair Both Unpaired Hole Cards
WILLNOTs : WILLs
.9917 : .0083
Reduce
.9917 / .0083 : .0083 / .0083
119.5 : 1

Open Backdoor 3-Straight Flush >>> Runner-Runner Straight Flush

With 3 suited and connected cards above **2** and below **K** in rank, there are **3** 2-card combinations that will make a Straight Flush:

Thus of a possible **1,081** 2-card combinations only **3 WILL** make the Straight Flush.

Knowing that with only **3** of over **1,000** possibilities you can approximate that the odds will be just over **300** to **1**:

Total Possibilities = 1,081
– WILLs = 3
WILLNOTs = 1,078

Odds of a RR Straight
WILLNOTs : WILLs
1,078 : 3
Reduce
1,078 / 3 : 3 / 3
359.3 : 1

Before the Last Card

After the Turn card has hit the board you have only 1 more card to come and only one more chance to improve your hand. But as we've said before: **hand improvement is not the only way to win the pot.**

With 1 card to come there are still 2 rounds of betting to come. In most limit games those rounds are at a higher stake than the first two.

With the higher cost to continue and the hand mostly made, players should have a precise strategy in order to remain involved.

At this juncture there are 3 ways you can win the pot:

1. Have the best hand and show it down on the River

2. Improve to the best hand on the last card and show it down on the River

3. Make a Bet, Raise or Check-Raise now and/or on the River that causes your opponent(s) to fold.

Knowing the likelihood that you can manipulate your opponent(s) into giving up the hand is a very important part of knowing your **Total Odds** of winning the pot. This makes knowing your opponents and their behavior an essential element to combine with the basic mathematical odds of improvement.

Odds of Improvement

This table shows the odds of improving certain common drawing hands on the last card.

Hand - Draw	Make on Last Card	Odds
Pair	2 Pair	2.8 : 1
Pair	Set	22 : 1
2 Pair	Full House	10.5 : 1
Set	Quads	45 : 1
Open Straight Draw	Straight	4.75 : 1
Gut-shot St Draw	Straight	10.75 : 1
4 Flush	Flush	4.11 : 1
4 Flush & Open Straight Draw	Straight or Flush	2.01 : 1
1 Pair & Open Straight Draw	Set or Straight	3.6 : 1
1 Pair & 4 Flush	Set or Flush	3.18 : 1
Open Straight Flush Draw	Straight Flush	22 : 1
2 Unpaired Cards	Pair Either Hole Card	6.67 : 1

The Calculations

With one card to come, the odds calculation for the various hands is straightforward. The only number that changes is the number of outs.

After the Turn card has hit the board there are 2 cards in your pocket, 4 on the board and **46** unseen cards:

Total Possibilities = 46

Total Possibilities − WILLs (Outs) = WILLNOTs

Odds = WILLNOTs : WILLs (Outs)

Pair >>> 2 Pair

There are **12** cards among the remaining 46 that will pair one of the 4 unpaired cards:

Total Possibilities = 46
− WILLs = 12
WILLNOTs = 34

Odds of Making 2 Pair with a Pair
WILLNOTs : WILLs
34 : 12
Reduce
34 / 12 : 12 / 12
2.8 : 1

Pair >>> Set

Whether the Pair is in the pocket or the player has paired one of his hole cards, there are still only **2** cards that will make the Set among the remaining 46. The odds of making a Set from a Pair on the Turn are

Total Possibilities = 46
− WILLs = 2
WILLNOTs = 44

Odds of Making a Set with a Pair
WILLNOTs : WILLs
44 : 2

Reduce

44 / 2 : 2 / 2

22 : 1

2 Pair >>> Full House

There are **4** cards among the remaining 46 that will make the Boat:

Total Possibilities = 46

– WILLs = 4

WILLNOTs = 42

Odds of Making a Full House with 2 Pair

WILLNOTs : WILLs

42 : 4

Reduce

42 / 4 : 4 / 4

10.5 : 1

Set >>> Quads

With a Set there is only **1** card among the remaining 46 that will make Quads:

Total Possibilities = 46

– WILLs = 1

WILLNOTs = 45

Odds of Making Quads with a Set

WILLNOTs : WILLs

45 : 1

Open Ended Straight Draw >>> a Straight

There are **8** cards, 4 on each end, that will make the Straight:

Total Possibilities = 46
− WILLs = 8
WILLNOTs = 38

Odds of Making a Straight with an Open Ended Draw
WILLNOTs : WILLs
38 : 8
Reduce
38 / 8 : 8 / 8
4.75 : 1

Gut-shot Straight Draw >>> Straight

With a gut-shot Straight draw there are **4** cards among the remaining 46 that will complete the draw:

Total Possibilities = 46
− WILLs = 4
WILLNOTs = 42

Odds of Making a Straight with a Gut-shot Draw
WILLNOTs : WILLs
43 : 4
Reduce
43 / 4 : 4 / 4
10.75 : 1

4 Flush >>> Flush

Again, whether there are 2 suited cards in the pocket or 1 suited card in the hole and 3 suited on the board, there are still **9** cards among the unseen 46 that will complete the Flush:

Total Possibilities = 46
– WILLs = 9
WILLNOTs = 37

Odds of Making a Flush with a 4-Flush
WILLNOTs : WILLs
37 : 9
Reduce
37 / 9 : 9 / 9
4.11 : 1

Open Straight Draw & 4 Flush >>> Straight or Flush

With an open ended Straight draw there are **8** outs to the Straight. With a 4 Flush there are **9** cards that will make the Flush. Because **2** of these cards will make both the Straight and the Flush there are **9 + 8 – 2 = 15** outs to either the Straight or the Flush:

Total Possibilities = 46
– WILLs = 15
WILLNOTs = 31

Odds of Making a Straight or a Flush
WILLNOTs : WILLs
31 : 15
Reduce
31 / 15 : 15 / 15
2.06 : 1

Pair & Open Straight Draw >>> Straight or Set

With a Pair and an open Straight draw there are **10** cards that will make either a Straight or a Set:

Total Possibilities = 46
– WILLs = 10
WILLNOTs = 36

Odds of Making a Set or Straight
with a Pair & Open Straight Draw
WILLNOTs : WILLs
36 : 10
Reduce
36 / 10 : 10 / 10
3.6 : 1

Pair & 4 Flush >>> Flush or Set

With a Pair and a Flush draw there are **11** cards that will make either a Flush or a Set:

Total Possibilities = 46
– WILLs = 11
WILLNOTs = 35

Odds of Making a Flush or Set
WILLNOTs : WILLs
35 : 11
Reduce
35 / 11 : 11 / 11
3.18 : 1

Open Ended Straight Flush Draw >>> Straight Flush

There are only **2** cards, either one of which will complete the Straight Flush draw:

Total Possibilities = 46
– WILLs = 2
WILLNOTs = 44

Odds of Making a Straight Flush
WILLNOTs : WILLs
44 : 2
Reduce
44 / 2 : 2 / 2
22 : 1

2 Unpaired Hole Cards >>> a Pair

With 2 unpaired hole cards there are **6** cards among the remaining unseen 46 cards that will make a Pair:

Total Possibilities = 46
– WILLs = 6
WILLNOTs = 40

Odds of Making a Pair w/2 Unpaired Hole Cards
WILLNOTs : WILLs
40 : 6
Reduce
40 / 6 : 6 / 6
6.67 : 1

The River Bet – Last Chance to Earn

All the cards are out — there is no chance of improving your hand. You can win now in one of two ways:

1. By having the best hand in a showdown

2. By using strategy to cause your opponent(s) to fold their cards.

Remember — your **Total Odds** of winning the pot are the odds that your hand will win in a showdown **combined** with the odds that you can manipulate your opponent(s) into giving up their stake in the pot.

It is at this point in the hand where the concepts of **Value-Bet** and **Value-Call** are most plainly demonstrated.

By this stage, three of the four rounds of betting have been completed. Pots are often large enough to offer positive money odds to a variety of seemingly risky propositions.

In a heads-up situation, if your best assessment is that you are a **3 to 1** dog to win the pot, but there are 10 bets in the pot, then there is an overlay to a call or a bet even without giving any consideration to whether or not your opponent(s) will call or fold.

There are 2 important points to remember on the River:

1. The greater your understanding of your opponents' likely reactions, the less you have to depend of the strength of your hand

2. The greater the strength of your hand, the less you have to worry about your opponents' reactions and the more you want them to call or raise.

The last round of betting can offer a great reward to a bold strategy.

With many bets in the pot and good knowledge of your opponent(s), a bet, a raise or even a check-raise with a busted draw can often pay as well as a completed draw.

The most valuable ability in all of Poker is the ability to win <u>without</u> having the best hand!

All players get the same number of good hands. It is the rare and often extremely successful player who can win pots with very strong hands and very weak hands alike.

4. Consolidated Odds Tables

In this chapter you will find a consolidation of the odds tables from both Chapter 4, *Odds in Texas Hold'em* and Chapter 6, *Odds in Omaha Hi-Lo*, presented in natural order by game and by round of betting.

These tables are provided as an easy-to-access, quick reference. While there are no explanations or expansions of the calculations in these tables, each item in each of these tables is fully explained and expanded in its respective game chapter.

Texas Hold'em Odds Tables

Starting Hands

This table displays the odds and the probabilities that you or any other specific player will find these starting hands in the hole on any specific hand.

Hand	Probability	Odds
Any Pair	5.9%	16 : 1
Any Specific Pair (AA)	.452%	220 : 1
Pair of Jacks or Better	1.8%	54.25 : 1
22 through TT	4.10%	23.6 : 1
Suited Connectors	2.10%	46.4 : 1
Suited Ace	3.62%	26.6 : 1
Any 2 Suited	23.53%	3.25 : 1
Any AK	1.2%	81.9 : 1
AKs	.30%	330.5 : 1
2 Unpaired Big Cards	7.2%	12.8 : 1
Any Ace	14.50%	5.9 : 1
Rags	81%	.23 : 1

*Note – Suited connectors calculated are **45** through **TJ**. **2 Big Cards** refers to any 2 unpaired cards **J** or higher.

Before the Flop

Given the starting hands/hole cards shown below, the following table displays the odds that those holdings will improve to the hands indicated.

Starting Hand	Hand to Flop	Odds
Pocket Pair	Set Only	8.28 : 1
Pocket Pair	Set or Boat	7.5 : 1
Pocket Pair	Boat	101.1 : 1
Pocket Pair	Quads	407.3 : 1
Pocket Pair	Set or Better	7.3 : 1
Pocket Pair	2 Pair	5.19 : 1
Suited Connectors	Flush Draw	8.14 : 1
Suited Connectors	Flush	117.8 : 1
Suited Connectors	Straight Draw	9.2 : 1
Suited Connectors	Straight	75.56 : 1
Suited Connectors	Straight Flush Draw	141 : 1
Suited Connectors	Straight Flush	4,899 : 1
2 Suited	Pair Either	2.45 : 1
2 Suited	Pair Both	48.49 : 1
2 to Royal Flush	Royal Flush Draw	138 : 1
2 to Royal Flush	Royal Flush	19,599 : 1
Ax Suited	Pair of Aces	5 : 1
Ax Suited	Ace Flush Draw	8.14 : 1
Ax Suited	Ace High Flush	117.8 : 1
2 Unpaired Cards	Pair a Certain Hole Card	5 : 1
2 Unpaired Cards	Pair Either Hole Card	2.45 : 1
2 Unpaired Cards	Pair Both Hole Cards	48.49 : 1
2 Unpaired Cards	A Set of Either Hole Card	73.24 : 1

After the Flop

Odds the Turn Card Will . . .

Given a certain flop, this table displays the odds of improvement with the Turn Card.

After The Flop	Hand to Turn	Odds
Pair	Set	22.5 : 1
4 Suited	Flush	4.22 : 1
Open Straight Draw	Straight	4.9 : 1
Gut-shot St Draw	Straight	10.75 : 1
Open St & 4 Flush	Straight or Flush	2.1 : 1
1 Pr & Open St Draw	Set or Straight	3.7 : 1
1 Pr & 4 Flush	Set or Flush	3.3 : 1
2 Pair	Full House	10.75 : 1
Set	Full House or Better	5.7 : 1
2 Unpaired Cards	Pair Either Hole Card	6.8 : 1

Odds with 2 Cards to Come

After the flop and with 2 cards to come, this table shows the odds that certain hands will improve on the Turn or on the River or on both the Turn and the River.

After The Flop	Hand After River Card	Odds
Pair	Set or Better	9.8 : 1
4 Suited	Flush	1.86 : 1
Open Straight Draw	Straight	2.18 : 1
Gut-shot Straight Draw	Straight	5.07 : 1
Open Straight & 4 Flush	Straight or Flush	.85 : 1
1 Pair & Open Straight Draw	Set or Straight	1.55 : 1
1 Pair & 4 Flush	Set or Flush	1.35 : 1
2 Pair	Full House or Better	5.1 : 1
Set	Full House or Better	1.95 : 1
2 Unpaired Cards	Pair or Better	3.1 : 1

Runner – Runner

The following table displays the odds of improving on both the Turn and the River to successfully complete the runner-runner draws indicated.

After The Flop	With Runner – Runner	Odds
1 Pair	Full House or Better	37.6 : 1
1 Pair	Quads	1,080 : 1
2 Pair	Quads	539.5 : 1
RR Open Straight Draw	Straight	21.52 : 1
RR 1 Gap Straight Draw	Straight	32.8 : 1

After The Flop	With Runner – Runner	Odds
RR 2 Gap Straight Draw	Straight	66.56 : 1
Backdoor Flush	Flush	23 : 1
2 Unpaired Cards	Set to Either Hole Card	179.2 : 1
2 Unpaired Cards	Pair Both Hole Cards	119.5 : 1
Open RR Straight Flush Draw	Straight Flush	359.3 : 1

Last Card (The River)

Below are the odds that these draws will improve on the River.

Hand - Draw	Make on Last Card	Odds
Pair	2 Pair	2.8 : 1
Pair	Set	22 : 1
2 Pair	Full House	10.5 : 1
Set	Quads	45 : 1
Open Straight Draw	Straight	4.75 : 1
Gut-shot St Draw	Straight	10.75 : 1
4 Flush	Flush	4.11 : 1
4 Flush & Open Straight Draw	Straight or Flush	2.01 : 1
1 Pair & Open Straight Draw	Set or Straight	3.6 : 1
1 Pair & 4 Flush	Set or Flush	3.18 : 1
Open Straight Flush Draw	Straight Flush	22 : 1
2 Unpaired Cards	Pair Either Hole Card	6.67 : 1

Omaha Hi-Lo Odds Tables

Starting Hands

The table below shows the odds and probabilities that you or any other specific player will find these certain starting cards in the hole on any particular hand.

Hand	Probability	Odds
AAXX	2.5%	39 : 1
A2XX	7.2%	12.8 : 1
A2 Suited to A	2.7%	36.1 : 1
A23X	1.2%	85.3 : 1
A23X Suited to A	.28%	351.5 : 1
AA2X	.42%	239 : 1
AA2X Single Suited to A	.21%	469 : 1
AA2X Double Suited	.05%	1879 : 1
AA23	.035%	2,819.1 : 1
AA23 Single Suited	.018%	5,639.1 : 1
AA23 Double Suited	.0044%	22,559.4 : 1
2 Pair	2.1%	47.2 : 1
Probably Not Playable	83.3%	.2 : 1

Before the Flop

This table shows the likelihood of improvement on the Flop in Omaha Hi-Lo.

Starting Hand	Hand to Flop	Odds
AA23	Nut Low Draw	.35 : 1
AA23	Nut Low	4.4 : 1
AA23	Wheel or Aces Full	65 : 1
AA23 Double Suited	Ace-High Flush Draw	2.4 : 1
AA23 Double Suited	Nut Low or Flush or Aces Full	3.7 : 1
AA23	Counterfeit A or 2 or 3	.92 : 1
AA2X	Nut Low Draw	.76 : 1
AA2X	Nut Low or Aces Full	3.3 : 1
AA2X Suited to A	Nut Low or Flush or Aces Full	3.2 : 1
A23X	Nut Low Draw	1.7 : 1
A23X	Nut Low	4.1 : 1
A23X Suited to A	Ace High Flush or Nut Low	3.9 : 1
A2XX	Nut Low Draw	.8 : 1
A2XX	Nut Low	3.5 : 1
A2XX	Wheel	269.3 : 1
A2XX Suited to A	Flush or Low	3.3 : 1
A2XX	Counterfeit A or 2	1.6 : 1
A3XX	Nut Low Draw	4.4 : 1
A3XX	Nut Low	26 : 1

Odds of Improvement with the Turn Card

This table reflects the odds of improvement with the Turn Card in Omaha Hi-Lo.

After The Flop	Hand to Turn	Odds
Low Draw	Low	1.8 : 1
Low Draw with Backup Low Card	Low	1.1 : 1
Wheel Draw	Wheel	10.3 : 1
Low Draw or Made Low	Counterfeit Low	6.5 : 1
Low Draw & Flush Draw	Low or Flush	1.1 : 1
Flush Draw	Flush	4 : 1
Open Straight Draw	Straight	4.6 : 1
Gut-shot Straight Draw	Straight	10.25 : 1
Open Straight & 4 Flush & Low Draw	Straight, Flush or Low	.67 : 1
2 Pair	Full House	10.3 : 1
Set	Full House or Quads	5.4 : 1
Made Straight or Flush	Pair the Board	4 : 1

Odds of Improvement with 2 Cards to Come

This table reflects the odds of improvement on either the Turn or the River or on both the Turn and the River in Omaha Hi-Lo.

After The Flop	Hand After River Card	Odds
Low Draw	Low	1.03 : 1
Low Draw With Backup Low Card	Low	.4 : 1
Wheel Draw	Wheel	4.8 : 1
Low Draw or Made Low	Counterfeit Low	3 : 1
Flush Draw	Flush	1.75 : 1
Low Draw & Flush Draw	Low or Flush	.4 : 1
Low Draw & Flush Draw	Low & Flush	8.2 : 1
Open Straight Draw	Straight	2.1 : 1
Gut-shot Straight Draw	Straight	4.8 : 1
2 Pair	Full House or Quads	4.6 : 1
Set	Full House or Quads	1.9 : 1
Made Straight or Flush	Pair the Board	1.4 : 1

Odds of Making Certain Runner-Runner Draws

This table shows the odds of improving on BOTH the Turn and the River to make the runner-runner draws noted below in Omaha Hi-Lo.

After The Flop	Runner – Runner	Odds
Back Door Low Draw	Low	5.2 : 1
Back Door Wheel Draw	Wheel	21.5 : 1
Back Door Flush Draw	Flush	26.5 : 1
Back Door Open Ended Straight Draw	Straight	19.6 : 1
4 Unpaired Hole Cards	Set	81.5 : 1

After The Flop	Runner – Runner	Odds
4 Unpaired Hole Cards	Pair 2 Hole Cards or Set	66.56 : 1
2 Pair in the Hole	Quads	494 : 1
Open Back Door Straight Flush Draw	Straight Flush	329 : 1

Odds of Improvement with Only the River Card to Come

This table reflects the odds of improvement with the River Card in Omaha Hi-Lo.

Hand – Draw	Make on Last Card	Odds
Low Draw	Low	1.75 : 1
Low Draw with Backup Low Card in the Hole	Low	1.1 : 1
Low Draw & Flush Draw	Low or Flush	1.1 : 1
2 Pair	Full House	10 : 1
Open Ended Straight Draw	Straight	4.5 : 1
Gut-shot Straight Draw	Straight	10 : 1
Flush Draw	Flush	3.9 : 1
Flush Draw & Open Straight Draw	Straight or Flush	1.9 : 1
1 Pair & Flush Draw	Set or Flush	3 : 1
Open Straight Flush Draw	Straight Flush	21 : 1
4 Unpaired Hole Cards	Pair Any Hole Card	2.7 : 1

5. Odds in Omaha Hi-Lo

In a 10-handed game of Omaha High or Omaha Hi-Lo, by the time the hand has been dealt, 45 of 52 cards are in play.

This means that in many cases, any nut hand that is possible has a very good chance of happening. For example, in Hold'em the odds of any player getting a pair of Aces in the hole is **220 : 1** but in Omaha it is **39 : 1**.

In a 10-handed Hold'em game some player at the table will, on average, have Aces in the hole only once in each 22 hands, but in Omaha, on average someone will have rockets once every 4 hands.

Conventional poker wisdom notes that Omaha, whether High or Hi-Lo, is as much about "nut-peddling" as it is about anything else. **The player should either have the best possible hand or be drawing to it.**

One pitfall of Hi-Lo is that you can have the nut Low hand and still lose money. Consider a pot that is contested heads-up. Each of you has invested $500 in a $1000 pot. You have the nut Low, but your opponent has a hand that matches your Low and beats your High. You win $250 and your opponent takes $750. You end the hand with a $250 loss. Being "quartered" or worse in Hi-Lo sends many naive players to the rail.

The idea in Hi-Lo is to scoop both sides of the pot and **NEVER** to chase with marginal hands in either direction. To demonstrate the power of the scoop, assume that you are in a 3-handed pot and each player has $400 invested (for a total pot of $1,200). If you hold the nut Low and you are not quartered, you win $600 and make a $200 profit. If you scoop, you win $1,200 with an $800 profit. **While your win is doubled, your profit is quadrupled.** That is the power of the scoop.

Starting Hands

Your first chance to gain an edge over your opponents is with your starting hands. The two keys to this edge are:

1. Play a higher quality starting hand than your opponents

2. Play the hand better than your opponents.

The strength of a hand matters more in split games than in non-split games. With both Low and High possibilities, calls are more likely and moves don't work as often.

In a 10-handed game, the players' hole cards represent over 3/4 of the deck. After the River card, 45 of 52 cards have been dealt to either the board or the players.

The significance of this almost complete distribution of the whole deck on every hand, is that whatever nut hands the board makes possible have a good chance of being held among the 40 cards at the table.

The table below shows the likelihood of you or any other specific player at the table receiving certain hands as hole cards in Omaha Hi-Lo.

Odds & Probabilities

Hand	Probability	Odds
AAXX	2.5%	39 : 1
A2XX	7.2%	12.8 : 1

Hand	Probability	Odds
A2 Suited to A	2.7%	36.1 : 1
A23X	1.2%	85.3 : 1
A23X Suited to A	.28%	351.5 : 1
AA2X	.42%	239 : 1
AA2X Single Suited to A	.21%	469 : 1
AA2X Double Suited	.05%	1879 : 1
AA23	.035%	2,819.1 : 1
AA23 Single Suited	.018%	5,639.1 : 1
AA23 Double Suited	.0044%	22,559.4 : 1
2 Pair	2.1%	47.2 : 1
Probably Not Playable	83.3%	.2 : 1

The Calculations

The odds and probabilities can be calculated at least 2 ways:

1. **Combinations** — By using the basic combinations as demonstrated in Chapter 1, *The Basic Calculations* and Chapter 4, *Odds in Texas Hold'em.*

2. **Probabilities** — The probability of any starting hand can be determined by multiplying the probabilities of the individual cards or the probabilities of the sets that make up the 4-card starting hand.

It will prove easier to combine sets of 2-card combinations for almost all of these calculations.

As demonstrated in Chapter 1, *The Basic Calculations* and throughout this book, any odds are easily converted into probabilities and *vice versa*.

A number that will prove useful throughout these calculations is the total number of possible 4-card combinations in a 52-card deck — **Comb(52, 4)**:

$$\frac{(52 * 51 * 50 * 49)}{(1 * 2 * 3 * 4)} = 270,725$$

There are **270,725** possible 4-card combinations from a deck of 52 cards.

AA

The odds of getting a pair of Aces in the hole in Omaha Hi-Lo underscores the dramatic difference between Hold'em, with 2 cards in the pocket, and Omaha Hi-Lo, with 4 cards in the pocket.

With the odds of pocket rockets in Hold'em at **220 : 1** and the odds of **AAxx** in the hole in Omaha Hi-Lo at **39 : 1**, you will see Aces in the hole more than 5 times as often in Omaha Hi-Lo as you will in Hold'em.

These odds can be calculated for exactly 2 Aces in the hole or for 2 or more Aces in the hole. The calculation that will serve you best is for exactly 2 Aces in the hole, because in Omaha Hi-Lo, 3 or more cards of the same rank in the hole is NOT a good thing. Omaha Hi-Lo is one of the very few places in all of poker where a hand containing 4 hidden Aces is a huge disappointment.

Start with **270,725** — the total possible 4-card combinations from a deck of 52 cards.

Find the number of 2-card combinations that contain a pair of Aces — **Comb(4, 2)**:

$$\frac{(4 * 3)}{(1 * 2)} = 6$$

The last number we need is the number of 2-card combinations that contain no Aces — **Comb(48, 2)**:

$$\frac{(48 * 47)}{(1 * 2)} = 1,128$$

To find the total number of 4-card combinations that contain **exactly** 2 Aces, multiply the number of 2-card combinations that **WILL** contain 2 Aces (**6**) by the number of 2-card combinations that **WILLNOT** contain any Aces (**1,128**):

$$6 * 1,128 = 6,768$$

The probability of 2 and only 2 Aces in the pocket in Omaha Hi-Lo expressed as a percentage is

$$\frac{\textbf{WILLs}}{\textbf{Total Possibilities}} * \textbf{100} = \textbf{Probability as \%}$$

$$\frac{\textbf{6.768}}{\textbf{270,725}} * \textbf{100} = \textbf{2.5\%}$$

The odds of exactly one pair of Aces in the hole in Omaha Hi-Lo are

Total Possibilities = 270,725
– WILLs = 6,768
WILLNOTs = 263,957

WILLNOTs (263,957) : WILLs (6,768)
263,957 : 6,768
Reduce
263,957 / 6,768 : 6,768 / 6,768
39 : 1

Any A2/AK

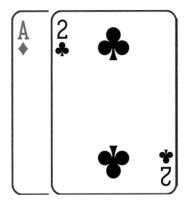 

The odds/probability of finding any **A2** or any **AK** in the hole in Omaha Hi-Lo are the same.

For this example we will calculate the odds of any **A2** in the hole in Omaha Hi-Lo.

The total number of 4-card combinations possible from a 52-card deck is **270,725**.

The number of 2-card combinations that contain **A2** is

$$4 * 4 = 16$$

The number of 2-card combinations possible from the remaining 50 cards in the deck, **including the possibility of one or more Aces or Deuces — Comb(50, 2)** is

$$\frac{(50 * 49)}{(1 * 2)} = 1,225$$

To find the number of 4-card combinations that will contain any **A2**, multiply the number of possible **A2** combinations (**16**) by the number of 2-card combinations possible from the remaining 50 cards in the deck:

$$16 * 1,225 = 19,600$$

The probability of **A2** in the pocket in Omaha Hi-Lo expressed as a percentage is

$$\frac{\textbf{WILLs}}{\textbf{Total Possibilities}} * 100 = \textbf{Probability as \%}$$

$$\frac{19,600}{270,725} * 100 = 7.2\%$$

The odds of **A2** in the hole in Omaha Hi-Lo are

Total Possibilities = 270,725
− WILLs = 19,600
WILLNOTs = 251,125

WILLNOTs (251,125) : WILLs (19,600)
251,125 : 19,600
Reduce
251,125 / 19,600 : 19,600 / 19,600
12.8 : 1

A2/AK Single Suited to the Ace

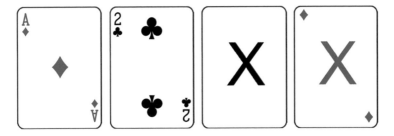

For the Flush possibilities, it doesn't matter whether the Deuce is suited to the Ace or not. It only matters that there is another card of the same suit as the Ace.

We will calculate the odds of having an **A2** with the Ace suited to either the Deuce or any of the other cards.

The total number of 4-card combinations possible from a 52-card deck is **270,725**.

The number of 2-card combinations that contain **A2,** including the possibility of a suited **A2** is

$$4 * 4 = 16$$

The number of 2-card combinations that contain one or more cards suited to the Ace is the number of cards, other than the Ace of that suit (**12**), multiplied by the remaining **50−12 = 38** cards in the deck:

$$12 * 38 = 456$$

To find the number of 4-card combinations that will contain an **A2** with the Ace suited to either the Deuce or 1 of the other cards in the hole, multiply the number of possible **A2** combinations (**16**) by the number of 2-card combinations that contain one card suited to the Ace (**456**):

$$16 * 456 = 7,296$$

The probability of **A2** with the Ace suited to either the Deuce or 1 of the other cards in the hole expressed as a percentage is:

$$\frac{\textbf{WILLs}}{\textbf{Total Possibilities}} * 100 = \textbf{Probability as \%}$$

$$\frac{7,296}{270,725} * 100 = 2.7\%$$

The odds of **A2** with the Ace suited to either the Deuce or 1 of the other cards in the hole are:

Total Possibilities = 270,725
– WILLs = 7,296
WILLNOTs = 263,429

WILLNOTs (263,429) : WILLs (7,296)
263,429 : 7,104
Reduce
263,429 / 7,104 : 7,104 / 7,104
36.1 : 1

Any A23

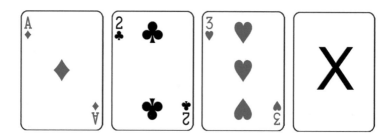

The total number of 4-card combinations possible from a 52-card deck is **270,725**.

The number of 3-card combinations that contain **A23** is:

$$4 * 4 * 4 = 64$$

The number of 4-card combinations that contain **A23**, including the possibility that one of the hole cards will be duplicated, is the number of 3-card combinations that contain **A23** (**64**) multiplied by the number of cards remaining in the deck, including the remaining Aces, Deuces and Threes (**49**):

$$64 * 49 = 3,136$$

Or you can multiply the number of 2-card combinations that are possible to make **A2** (**16**) by the number of 2-card combinations that contain at least one **3** (**4 * 49 = 196**):

$$16 * 196 = 3,136$$

The probability of **A23** in the pocket expressed as a percentage is:

$$\frac{\textbf{WILLs}}{\textbf{Total Possibilities}} * \textbf{100} = \textbf{Probability as \%}$$

$$\frac{\textbf{3,136}}{\textbf{270,725}} * \textbf{100} = \textbf{1.2\%}$$

The odds of **A23** in the pocket are:

Total Possibilities = 270,725
– WILLs = 3,136
WILLNOTs = 267,589

WILLNOTs (267,589) : WILLs (3,136)
267,589 : 3,136
Reduce
267,589 / 3,136 : 3,136 / 3,136
85.3 : 1

A23 Suited to the Ace

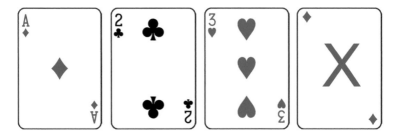

For the Flush possibilities, it doesn't matter whether the **2** or the **3** is suited to the Ace — it only matters that there is another card of the same suit as the Ace.

For this example we will calculate the odds of having an **A23** with the Ace suited to the Deuce, the Trey or the fourth card in the hole.

This calculation includes the possibility that more than one of the hole cards will be suited to the Ace.

We know that the total number of 4-card combinations possible from a 52-card deck is **270,725**.

The number of 3-card combinations that contain **A23**, including the possibility of a suited **A2** or suited **A3** is:

$$4 * 4 * 4 = 64$$

To find the number of 4-card combinations that will contain an **A23** with the Ace suited to the Deuce, the Trey, or the other card in the hole, multiply the number of possible **A23** combinations (**64**) by **12**, the number of remaining cards of the same suit as the Ace:

$$64 * 12 = 768$$

The probability of **A23** with the Ace suited to the Deuce, the Trey or the other card in the hole expressed as a percentage is

$$\frac{\text{WILLs}}{\text{Total Possibilities}} * 100 = \text{Probability as \%}$$

$$\frac{768}{270,725} * 100 = .28\%$$

Both of these calculations include the possibility of some duplication if the Ace is suited to either or both the Deuce and the Trey.

The odds of **A23** with the Ace suited to the Deuce, the Trey or the other card in the hole are:

Total Possibilities = 270,725
− WILLs = 768
WILLNOTs = 269,957

WILLNOTs (269,957) : WILLs (768)
269,957 : 768
Reduce
269,957 / 768 : 768/ 768
351.5 : 1

Any AA2X

The total number of 4-card combinations possible from a 52-card deck is **270,725**.

There are **6** 2-card combinations that can contain a pair of Aces.

To find the number of 4-card combinations that can contain **AA2x**, multiply the number of possible pairs of Aces (**6**) by the number of 2-card combinations that can contain at least one Deuce and no Aces (**4 * 47 = 188**). The number of 4-card combinations that can contain one pair of Aces and one or more Deuces is:

$$6 * 188 = 1,128$$

The probability of **AA2x** in the hole expressed as a percentage is

$$\frac{\textbf{WILLs}}{\textbf{Total Possibilities}} * \textbf{100} = \textbf{Probability as \%}$$

$$\frac{\textbf{1.128}}{\textbf{270,725}} * \textbf{100} = \textbf{.42\%}$$

To find the odds of **AA2x** in the hole:

Total Possibilities = 270,725
– WILLs = 1,128
WILLNOTs = 269,597

WILLNOTs (269,597) : WILLs (1,128)
269,597 : 1,128
Reduce
269,597 / 1,128 : 1,128/ 1,128
239 : 1

AA2X Single Suited to A

For the Flush possibilities it doesn't matter whether the Deuce or the fourth card is suited to one of the Aces. It only matters that one of those cards is suited to one of the Aces.

In this example we will calculate the odds of having an **AA2** with one of the Aces suited to either the Deuce or the fourth card in the hole.

This includes the possibility that more than one of the hole cards will be suited to either Ace.

The total number of 4-card combinations possible from a 52-card deck is **270,725**.

There are **6** 2-card combinations that can contain a pair of Aces.

To find the number of 4-card combinations that can contain **AA2x** single suited, multiply the number of possible pairs of Aces (**6**), by the number of 2-card combinations that can contain at least one Deuce and one of the 24 remaining cards of the same suit as either Ace (**4 ∗ 24 = 96**). The number of 4-card combinations that can contain one pair of Aces, one or more Deuces and one card suited to either of the Aces is:

$$6 * 96 = 576$$

The probability of **AA2x** single suited in the hole expressed as a percentage:

$$\frac{\text{WILLs}}{\text{Total Possibilities}} * 100 = \text{Probability as \%}$$

$$\frac{576}{270,725} * 100 = .21\%$$

The odds of **AA2x** are

Total Possibilities = 270,725
− WILLs = 576
WILLNOTs = 270,149

WILLNOTs (270,149) : WILLs (576)
270,149 : 576
Reduce
270,149 / 576 : 576 / 576
469 : 1

AA2X Double Suited

Here we will calculate the odds of having an **AA2** with both Aces suited, one Ace to the Deuce and the second Ace suited the 4th card in the hole.

The total number of 4-card combinations possible from a 52-card deck is **270,725**.

There are **6** 2-card combinations that can contain a pair of Aces.

To find the number of 4-card combinations that can contain **AA2x** double suited, multiply the number of possible pairs of Aces (**6**), by the number of 2-card combinations that can contain at least one Deuce that is suited to one of the Aces and to one of the 12 remaining cards of the same suit as the other Ace (**2 * 12 = 24**). The number of 4-card combinations that can contain one pair of Aces and one or more Deuces and is double suited to each of the Aces is

$$6 * 24 = 144$$

The probability of **AA2x** double suited in the hole expressed as a percentage:

$$\frac{\textbf{WILLs}}{\textbf{Total Possibilities}} * 100 = \textbf{Probability as \%}$$

$$\frac{144}{270,725} * 100 = .05\%$$

The odds of **AA2x** double suited in the hole are

Total Possibilities = 270,725
– WILLs = 144
WILLNOTs = 270,581

WILLNOTs (270,581) : WILLs (144)
270,581 : 144
Reduce
270,581 / 144 : 144/ 144
1,879 : 1

Any AA23/AA2K

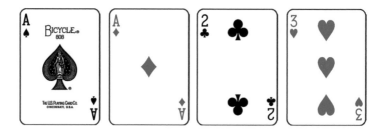

The total number of 4-card combinations possible from a 52-card deck is **270,725**.

There are **6** 2-card combinations that can contain **AA**.

There are **4 * 4 = 16** 2-card combinations that contain **23**.

The number of 4-card combinations that can contain **AA23** is the number of possible 2-card combinations that contain 2 Aces (**6**), multiplied by the number of possible 2-card combinations that contain **23** (**16**):

$$6 * 16 = 96$$

The probability of **AA23** in the pocket expressed as a percentage is:

$$\frac{\text{WILLs}}{\text{Total Possibilities}} * 100 = \text{Probability as \%}$$

$$\frac{96}{270,725} * 100 = .035\%$$

The odds of **A23** in the pocket are:

Total Possibilities = 270,725
– WILLs = 96
WILLNOTs = 270,629

WILLNOTs (270,629) : WILLs (96)
270,629 : 96
Reduce
270,629 / 96 : 96 / 96
2,819.1 : 1

AA23/AA2K Single Suited

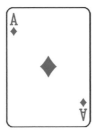

We will calculate the odds of having an **AA23** single suited to either Ace in the hole.

The total number of 4-card combinations possible from a 52-card deck is **270,725**.

There are **6** 2-card combinations that can contain a pair of Aces.

To find the number of 4-card combinations that can contain **AA23** single suited, multiply the number of possible pairs of Aces (**6**), by the number of 2-card combinations that can contain at least one Deuce, one Trey and with either the **2** or the **3** suited to one of the Aces (**2 ∗ 4 = 8**). The number of 4-card combinations that can contain **AA23** single suited is

$$6 * 8 = 48$$

The probability of **AA23** single suited in the hole expressed as a percentage is

$$\frac{\text{WILLs}}{\text{Total Possibilities}} * 100 = \text{Probability as \%}$$

$$\frac{48}{270,725} * 100 = .018\%$$

The odds of **AA23** single suited in the hole are:

Total Possibilities = 270,725
− WILLs = 48
WILLNOTs = 270,677

WILLNOTs (270,677) : WILLs (48)
270,677 : 48
Reduce
270,677 / 48 : 48 / 48
5,639.1 : 1

AA23/AA2K Double Suited

Here we will calculate the odds of having an **AA23** double suited to both Aces in the hole.

The total number of 4-card combinations possible from a 52-card deck is **270,725**.

There are **6** 2-card combinations that can contain a pair of Aces.

There are only **2** 2-card combinations of **23** that will be suited to both Aces. The number of 4-card combinations that can contain **AA23** double suited is

$$6 * 2 = 12$$

The probability of **AA23** double suited in the hole expressed as a percentage is

$$\frac{\textbf{WILLs}}{\textbf{Total Possibilities}} * 100 = \textbf{Probability as \%}$$

$$\frac{12}{270,725} * 100 = .0044\%$$

The odds of **AA23** double suited in the hole are

Total Possibilities = 270,725
– WILLs = 12
WILLNOTs = 270,713

WILLNOTs (270,713) : WILLs (12)
270,713 : 12
Reduce
270,713 / 12 : 12 / 12
22,559.4 : 1

2 Pair

We will calculate the odds of finding any 2 pair in the hole.

The total number of 4-card combinations possible from a 52-card deck is **270,725**.

There are **78** possible 2-card combinations that can contain a pair.

To find the number of 4-card combinations that will contain 2 pair, multiply the number of possible pair combinations (**78**), by the number of possible pairs excluding pairs of the rank of the first pair (**72**):

$$78 * 72 = 5{,}616$$

The probability of 2 pair in the pocket expressed as a percentage is:

$$\frac{\textbf{WILLs}}{\textbf{Total Possibilities}} * 100 = \textbf{Probability as \%}$$

$$\frac{\textbf{5,616}}{\textbf{270,725}} * 100 = \textbf{2.1\%}$$

The odds of **A2** in the hole are:

$$\textbf{Total Possibilities} = \textbf{270,725}$$
$$-\textbf{WILLs} = \textbf{5,616}$$
$$\textbf{WILLNOTs} = \textbf{265,109}$$

$$\textbf{WILLNOTs (265,109) : WILLs (5,616)}$$
$$\textbf{265,109 : 5,616}$$
$$\textbf{Reduce}$$
$$\textbf{265,109 / 5,616 : 5,616 / 5,616}$$
$$\textbf{47.2 : 1}$$

Probably Not Playable

With the three *caveats* that

1. **not all of the hands listed below are always playable;**

2. **there is some <u>duplication</u> among the hands in the table below; and**

3. **several usually playable hands are not listed;**

the following gives a fair approximation of the odds of rags in the hole.

Because we are trying to find the odds of getting a non-playable hand, the number of **playable** hands represents the **WILLNOTs**:

Hand	Number of Possible Hands
AAXX	6,768
A2XX	19,600
A2 Suited to A	7,104
A23X	3,136
A23X Suited to A	768
AA2X	1,128
AA2X Single Suited to A	576
AA2X Double Suited	144
AA23	96
AA23 Single Suited	48
AA23 Double Suited	12
2 Pair	5,616
Total Playable Hands	44,996

The probability of getting a non-playable hand expressed as a percentage is:

Total Possibilities = 270,725
− WILLNOTs = 44,996
WILLs = 225,729

$$\frac{\textbf{WILLs}}{\textbf{Total Possibilities}} * \textbf{100} = \textbf{Probability as \%}$$

$$\frac{\textbf{225,729}}{\textbf{270,725}} * \textbf{100} = \textbf{83.3\%}$$

To find the odds of getting a non-playable hand in the hole:

WILLNOTs (44,996) : WILLs (225,729)
44,996 : 225,729
Reduce
44,996 / 225,729 : 225,729 / 225,729
.2 : 1

Before the Flop

In both Omaha High and Omaha Hi-Lo, a primary consideration is the odds of improving the hand on the Flop. In Omaha Hi-Lo, an additional concern is the likelihood that a Low hand or Low draw will be counterfeited.

Second-nut Lows, second-nut Flushes and "dummy" end Straights are all hands that can be very expensive for two reasons:

1. They are second-nut — they have a diminished chance of winning

2. They are mostly unidirectional hands — they have virtually no chance of scooping.

The tables and calculations below demonstrate the odds and probabilities that

1. Certain starting low hands will improve into Low draws on the Flop

2. Certain starting high hands will improve into High draws on the Flop

3. Either low or high starting hands will evolve into made hands on the Flop.

Certain starting low hands will be **counterfeited** on the Flop.

Omaha Hi-Lo – Before the Flop

Starting Hand	Hand to Flop	Odds
AA23	Nut Low Draw	.35 : 1
AA23	Nut Low	4.4 : 1
AA23	Wheel or Aces Full	65 : 1
AA23 Double Suited	Ace-High Flush Draw	2.4 : 1
AA23 Double Suited	Nut Low or Flush or Aces Full	3.7 : 1
AA23	Counterfeit A or 2 or 3	.92 : 1
AA2X	Nut Low Draw	.76 : 1
AA2X	Nut Low or Aces Full	3.3 : 1
AA2X Suited to A	Nut Low or Flush or Aces Full	3.2 : 1
A23X	Nut Low Draw	1.7 : 1
A23X	Nut Low	4.1 : 1
A23X Suited to A	Ace High Flush or Nut Low	3.9 : 1
A2XX	Nut Low Draw	.8 : 1
A2XX	Nut Low	3.5 : 1
A2XX	Wheel	269.3 : 1
A2XX Suited to A	Flush or Low	3.3 : 1
A2XX	Counterfeit A or 2	1.6 : 1
A3XX	Nut Low Draw	4.4 : 1
A3XX	Nut Low	26 : 1

The Calculations

These calculations will use both the comparison of sets and the combination of probabilities.

The total number of possible different 3-card Flops will be used throughout these calculations. With your 4 hole cards, the total possible 3-card Flops from the remaining 48 cards is **Comb(48, 3)**:

$$\frac{(48 * 47 * 46)}{(1 * 2 * 3)} = 17{,}296$$

Another number that will useful in these calculations is the number of 2-card combinations possible from the remaining unseen 48 cards — **Comb(48,2)**:

$$\frac{(48 * 47)}{(1 * 2)} = 1{,}128$$

AA23

Starting Hand	Hand to Flop	Odds
AA23	Nut Low Draw	.35 : 1
AA23	Nut Low	4.4 : 1
AA23	Wheel or Aces Full	65 : 1
AA23 Double Suited	Nut Flush Draw	2.4 : 1
AA23 Double Suited	Nut Low or Flush or Aces Full	3.7 : 1
AA23	Counterfeit A or 2 or 3	.92 : 1

AA23 is probably the best starting hand in Omaha Hi-Lo. The odds of its improvement in either direction are better

than for almost any other hand. You will flop a nut Low draw over 74% of the time with this hand.

AA23 >>> Nut Low Draw

To flop a nut Low draw with this hand, the Flop needs to contain any 2-card combination that has 2 unpaired low cards, exactly one of which may be of the same rank as any one of the hole cards.

This calculation includes the possibility of flopping a made nut Low.

There are 32 cards in the deck below the rank of **9**, and 4 of them are in the hole. The number of 2-card combinations that contain 2 cards below the rank of **9** is **Comb(28, 2)**:

$$\frac{(28 * 27)}{(1 * 2)} = 378$$

This number includes **30** pairs from the ranks **4** through **8**, **1** pair of Aces, **3** pairs of Deuces, and **3** pairs of Treys, for a total of **37** pairs that must be subtracted from **378**:

$$378 - 37 = 341$$

From this, subtract the unpaired 2-card combinations both of which match hole cards: **6 A2** combinations, plus **6 A3** combinations, plus **9 2?3** combinations, for a total of **21** combinations that match 2 hole cards:

$$341 - 21 = 320$$

To include the possibility of flopping a made Low hand, multiply this number (**320**), by the number of remaining cards in the deck (48), minus the 2 Flop cards that make the

Low hand (2), minus the 6 cards that will pair either of the first 2 Flop cards (**48 – 2 – 6 = 40**):

$$320 * 40 = 12,800$$

There are **12,800** 3-card Flops the **WILL** make a nut Low draw or better:

Total Possibilities = 17,296
– WILLs = 12,800
WILLNOTs = 4,496

Odds of Flopping Nut Low Draw or Better
WILLNOTs : WILLs
4,496 : 12,800
Reduce
4,496 / 12,800 : 12,800 / 12,800
.35 : 1

To express these odds as a probability:

$$\frac{\textbf{WILLs}}{\textbf{Total Possibilities}} * 100 = \textbf{Probability as \%}$$

$$\frac{\textbf{12,800}}{\textbf{17,296}} * 100 = 74\%$$

AA23 >>> Nut Low

To flop a nut Low with this hand, the Flop must contain 2 unpaired low cards that do not match any of the hole cards, plus a third low card that may match any of the hole cards, but may not match either of the first two cards flopped.

First you need to calculate the number of 2-card combinations that will make the nut Low draw without counterfeiting any of the hole cards.

There are **20** cards ranked **8** or lower that do not match any of the hole cards. The number of 2-card combinations — **Comb(20, 2)**:

$$\frac{(20 * 19)}{(1 * 2)} = 190$$

From the ranks **4** through **8,** there are 30 possible pairs that must be subtracted from the total to indicate the number of 2-card combinations that will make the draw:

$$190 - 30 = 160$$

While the third card may counterfeit one of the hole cards and still produce a nut Low, it must not be a card above the rank of **8** and it must not pair one of the other 2 board cards. Thus there are **46 – 20** (number of cards **9** through **K**) — **6** (number of cards that will pair either of the board cards) = **20** possibilities for the third card. The number of 3-card Flops that **WILL** make a nut Low to this hand:

$$160 * 20 = 3,200$$

With a total of **17,296** possible 3-card Flops, the odds of flopping a nut Low or a Set of Aces:

Total Possibilities = 17,296
– WILLs = 3,200
WILLNOTs = 14,096

Odds of Flopping Nut Low or AAA
WILLNOTs : WILLs
14,096 : 3,200
Reduce
14,096 / 3,200 : 3,200 / 3,200
4.4 : 1

AA23 >>> Wheel or Aces Full

To flop a Wheel (**A2345**) with this hand, there must be both a **4** and a **5** on the board as well as a counterfeit to one of the hole cards.

There are **4** * **4** = **16** possible **45** combinations multiplied by one of the **8** cards that will counterfeit one of the hole cards:

$$16 * 8 = 128$$

128 3-card combinations will make a Wheel to this hand.

To flop Aces Full with this hand, one of the Flop cards must be an Ace and the other 2 must be paired. Remaining unseen are **1** pair of Aces, **3** pairs of Deuces, **3** pairs of Treys, and **60** pairs from the ten ranks **4** through **K**. With **2** unseen Aces, the number of 3-card Flops that will make Aces Full is:

$$67 * 2 = 134$$

128 + **134** = **262** 3-card Flops **WILL** make Aces Full or a Wheel to this hand:

Total Possibilities = 17,296
– WILLs = 262
WILLNOTs = 17,034

Odds of Flopping a Wheel or Aces Full
WILLNOTs : WILLs
17,034 : 262
Reduce
17,034 / 262 : 262 / 262
65 : 1

AA23 Double Suited >>> Nut Flush Draw

To flop a nut Flush draw, the Flop need only contain 2 cards of either of the 2 suits that match the hole cards. With **11** unseen cards of each of **2** suits, the number of possible 2-card combinations that will flop the nut Flush draw is **Comb(11, 2) * 2**:

$$\frac{(11 * 10)}{(1 * 2)} * 2 = 110$$

Multiply this by **46** (48 unseen cards minus 2), to get **46 * 110 = 5,060** possible Flops that will produce a nut Flush draw or better to this hand.

This calculation includes the possibility of flopping a made nut Low.

Total Possibilities = 17,296
– WILLs = 5,060
WILLNOTs = 12,236

Odds of Flopping a Nut Flush Draw or Better
WILLNOTs : WILLs
12,236 : 5,060
Reduce
12,236 / 5,060 : 5,060 / 5,060
2.4 : 1

AA23 Double Suited >>> Flush or Nut Low or Aces Full

With the *caveat* that there is some small overlap, we will find the number of Flops that will fill each of these hands separately, then add them to find the odds of making any of these three **very** strong hands.

From the calculations above, we know that there are **3,200** Flops that will make a nut Low.

To flop an Ace-High Flush, the Flop must contain 3 cards from either of 2 suits — **Comb(11, 3) * 2**:

$$\frac{(11 * 10 * 9)}{(1 * 2 * 3)} * 2 = 330$$

Also from the work above, we know that there are **134** Flops that will make Aces Full.

There are **3,664** Flops that will make a nut Low, Ace High Flush or Aces Full to this hand:

Nut Low = 3,200
Ace High Flush = 330
Aces Full = 134
3,664

Odds of Flopping a Nut Low, Ace High Flush
or Aces Full
WILLNOTs : WILLs
13,632 : 3,664
Reduce
13,632 / 3,664 : 3,664/ 3,664
3.7 : 1

AA23 >>> Counterfeited A or 2 or 3

To find the odds of a counterfeited **A, 2** or **3** at least once on the Flop, multiply the number of 2-card combinations possible from the 48 unseen cards by the number of remaining **A**'s, **2**'s and **3**'s — **Comb(48, 2) * 8**:

$$\frac{(48 * 47)}{(1 * 2)} * 8 = 9,024$$

Total Possibilities = 17,296
– WILLs = 9,024
WILLNOTs = 8,272

Odds of Flopping a Counterfeited A or 2 or 3
WILLNOTs : WILLs
8,272 : 9,024
Reduce
8,272 / 9,024 : 9,024/ 9,024
.92 : 1

AA2X >>> Nut Low Draw

To flop a nut Low draw with this hand, the Flop needs to contain 2 unpaired cards of the ranks **3** through **8** (no card may counterfeit either the Ace or the Deuce in the hole).

This calculation includes the possibility of flopping a made nut Low.

There are **Comb(24, 2)** 2-card combinations from the ranks **3** through **8** and **36** pairs:

$$\frac{(24 * 23)}{(1 * 2)} = 276$$

276 – 36 = 240 2-card combinations that will flop the nut Low draw. Multiply this by all of the remaining cards minus those that will counterfeit either the Ace or Deuce in the hole, **41**:

$$240 * 41 = 9,840$$

With **9,840** 3-card combinations that will flop the nut Low draw to this hand, the odds are

Total Possibilities = 17,296
– WILLs = 9,840
WILLNOTs = 7,456

Odds of Flopping a Nut Low Draw
WILLNOTs : WILLs
7,456 : 9,840
Reduce
7,456 / 9,840 : 9,840 / 9,840
.76 : 1

AA2X >>> Nut Low or Aces Full

To flop a nut Low with this hand, the board must contain 3 unpaired cards ranked **3** through **8**. There are 6 ranks and thus **Comb(24, 2)** 2-card combinations minus the 36 possible pairs among these ranks:

$$\frac{(24 * 23)}{(1 * 2)} = 276$$

276 – 36 = 240 is the number of 2-card combinations that will flop the nut Low draw. This is multiplied by all of the remaining cards ranked **3** through **8** that are not represented by the first 2 cards of the Flop (**16**), to produce the number of 3-card combinations that will make a nut Low to this hand:

$$240 * 16 = 3,840$$

To flop Aces Full, the board must contain an Ace and **any pair**. With this hand there are 66 unseen pairs ranked **3** through **K**, plus 1 pair of Aces and 3 pairs of Deuces = 70 possible pairs that might appear on the board. This is multiplied by the number of unseen Aces (2), to produce the number of possible 3-card flops that will make Aces Full or better to this hand:

$$70 * 2 = 140$$

This calculation includes the possibility of flopping an AA to produce QUADS.

There are **3,840 + 140 = 3,980** possible 3-card Flops that will make a nut Low hand or Aces Full or better to this hand. The odds of this Flop are

Total Possibilities = 17,296
− WILLs = 3,980
WILLNOTs = 13,316

Odds of Flopping a Nut Low or Aces Full
WILLNOTs : WILLs
13,316 : 3,980
Reduce
13,316 / 3,980 : 3,980 / 3,980
3.3 : 1

AA2X − Single Suited >>> Ace High Flush or Nut Low or Aces Full

There are **3,980** 3-card Flops that will make either a nut Low hand or Aces Full. To complete this calculation we must also find the number of 3-card Flops all of the same suit as the suited Ace — **Comb(11, 3)**:

$$\frac{(11 * 10 * 9)}{(1 * 2 * 3)} = 165$$

There are **3,980 + 165 = 4,145** 3-card Flops that will make an Ace High Flush, a nut Low or Aces Full to this hand. The odds against one of these flops are:

Total Possibilities = 17,296
− WILLs = 4,145
WILLNOTs = 13,151

**Odds of Flopping an Ace High Flush or Nut Low
or Aces Full**
WILLNOTs : WILLs
13,151 : 4,145
Reduce
13,151 / 4,145 : 4,145 / 4,145
3.2 : 1

A23X >>> Nut Low Draw

To flop a nut Low draw with this hand, the Flop needs to contain 2 unpaired cards below the rank of **9** with 1 and only 1 that may pair the **A**, **2** or **3** in the hole.

This calculation includes the possibility of flopping a made nut Low.

There are **Comb(20, 2)** 2-card combinations from the ranks **4** through **8** and **30** pairs:

$$\frac{(20 * 19)}{(1 * 2)} = 190$$

190 − 30 = 160 is the number of 2-card combinations that will flop a nut Low draw to this hand. This is multiplied by all of the remaining cards minus the 6 that might counterfeit one of the Flop cards, **40**:

$$160 * 40 = 6,400$$

With **6,400** 3-card combinations that will flop the nut Low draw to this hand, the odds are:

**Total Possibilities = 17,296
− WILLs = 6,400
WILLNOTs = 10,896**

Odds of Flopping a Nut Low Draw
WILLNOTs : WILLs
10,896 : 6,400
Reduce
10,896 / 6,400 : 6,400 / 6,400
1.7 : 1

A23X >>> Nut Low

To flop a nut Low with this hand the Flop needs to contain 3 unpaired cards below the rank of **9** with one and only one that may pair the **A**, **2** or **3** in the hole.

There are **Comb(20, 2)** 2-card combinations from the ranks **4** through **8** and **30** pairs:

$$\frac{(20 * 19)}{(1 * 2)} = 190$$

190 − 30 = 160 is the number of 2-card combinations that will flop a nut Low draw to this hand. This is multiplied by all of the remaining low cards minus the 6 that might counterfeit one of the flop cards (**21**):

$$160 * 21 = 3,360$$

With **3,360** 3-card combinations that will flop the nut Low to this hand, the odds are

Total Possibilities = 17,296
− WILLs = 3,360
WILLNOTs = 13,936

Odds of Flopping a Nut Low
WILLNOTs : WILLs
13,936 : 3,360

Reduce
13,936 / 3,360 : 3,360/ 3,360
4.1 : 1

A23X Single Suited >>> Nut Low or Ace High Flush

There are **3,360** Flops that will make a nut low.

To flop an Ace-High Flush, the Flop must contain 3 cards from the same suit as the suited Ace — **Comb(11, 3)**:

$$\frac{(11 * 10 * 9)}{(1 * 2 * 3)} = 165$$

3,360 + 165 = 3,525 is the number of 3-card combinations that will flop a Nut Low or an Ace-High Flush to this hand:

Total Possibilities = 17,296
– WILLs = 3,525
WILLNOTs = 13,771

Odds of Flopping a Nut Low or Ace High Flush
WILLNOTs : WILLs
13,771 : 3,525
Reduce
13,771 / 3,525 : 3,525 / 3,525
3.9 : 1

A2XX >>> Nut Low Draw

To flop a nut Low draw with this hand, the Flop must contain 2 unpaired cards of the ranks **3** through **8** with no card that counterfeits either the Ace or the Deuce in the hole.

This calculation includes the possibility of flopping a <u>made</u> nut Low.

There are **Comb(24, 2)** 2-card combinations from the ranks **3** through **8** and 36 pairs:

$$\frac{(24 * 23)}{(1 * 2)} = 276$$

276 – 36 = 240 is the number of 2-card combinations that will flop the nut Low draw. This is multiplied by all of the remaining cards minus those that will counterfeit either the Ace or Deuce in the hole (**40**):

$$240 * 40 = 9{,}600$$

With **9,600** 3-card combinations that will flop the nut Low draw to this hand, the odds are

Total Possibilities = 17,296
– WILLs = 9,600
WILLNOTs = 7,696

Odds of Flopping Nut Low Draw
WILLNOTs : WILLs
7,696 : 9,600
Reduce
7,696 / 9,600 : 9,600 / 9,600
.8 : 1

A2XX >>> Nut Low

To flop a nut Low with this hand, the board must contain 3 unpaired cards ranked **3** through **8**. There are six ranks and thus **Comb(24, 2)** 2-card combinations minus the 36 possible pairs among these ranks:

186

$$\frac{(24 * 23)}{(1 * 2)} = 276$$

276 − 36 = 240 is the number of 2-card combinations that will flop the nut Low draw. This is multiplied by all of the remaining cards ranked **3** through **8** that are not represented by the first 2 cards of the Flop (**16**), to produce the number of 3-card combinations that will make a nut Low to this hand:

240 * 16 = 3,840

Total Possibilities = 17,296
− WILLs = 3,840
WILLNOTs = 13,456

Odds of Flopping Nut Low
WILLNOTs : WILLs
13,456 : 3,840
Reduce
13,456 / 3,840 : 3,840 / 3,840
3.5 : 1

A2XX >>> A Wheel (A2345)

To flop a Wheel with this hand, the board will need to show **345**.

There are **4 * 4 * 4 = 64** 3-card combinations that contain **345** and will make a wheel.

Total Possibilities = 17,296
− WILLs = 64
WILLNOTs = 17,232

Odds of Flopping a Wheel
WILLNOTs : WILLs
17,232 : 64

Reduce
17,232 / 64 : 64 / 64
269.3 : 1

A2XX Suited Ace >>> Nut Low or Ace Flush

There are **3,840** Flops that will make a nut Low to this hand.

To flop an Ace-High Flush the Flop must contain 3 cards from the same suit as the Ace — **Comb(11, 3)**:

$$\frac{(11 * 10 * 9)}{(1 * 2 * 3)} = 165$$

3,840 + 165 = 4,005 is the number of 3-card combinations that will flop a nut Low or an Ace-High Flush to this hand:

Total Possibilities = 17,296
– WILLs = 4,005
WILLNOTs = 13,291

Odds of Flopping a Nut Low or an Ace High Flush
WILLNOTs : WILLs
13,291 : 4,005
Reduce
13,291 / 4,005 : 4,005 / 4,005
3.3 : 1

A2XX >>> Counterfeit Either or Both the Ace or the Deuce at Least Once

To find the odds of a counterfeited **A** or **2** at least once on the Flop, multiply the number of 2-card combinations possible from the **48** unseen cards by the number of remaining Aces and Deuces — **Comb(48, 2)** * **6**:

$$\frac{(48 * 47)}{(1 * 2)} * 6 = 6{,}768$$

Total Possibilities = 17,296
– WILLs = 6,768
WILLNOTs = 10,528

Odds of Counterfeiting A or 2
WILLNOTs : WILLs
10,528 : 6,768
Reduce
10,528 / 6,768 : 6,768 / 6,768
1.6 : 1

A3XX >>> Nut Low Draw

To flop a nut Low draw with this hand, the Flop needs to contain at least one Deuce and one other low card other than an Ace or Trey.

This calculation includes the possibility of flopping a made nut Low.

There are **20** cards ranked **4** through **8** plus **4** Deuces. Thus there are **20 * 4 = 80** 2-card combinations that contain a Deuce and one other low card. After these 2 cards are dealt, there are **40** remaining cards that will not counterfeit the Ace or the Trey. **80 * 40 = 3,200** 3-card combinations will flop a nut Low draw to this hand:

Total Possibilities = 17,296
– WILLs = 3,200
WILLNOTs = 14,096

Odds of Flopping Nut Low Draw
WILLNOTs : WILLs
14,096 : 3,200

Reduce
14,096 / 3,200 : 3,200 / 3,200
4.4 : 1

A3XX >>> Nut Low

To flop the Nut Low with this hand, the board must contain 2 unpaired cards ranked **4** through **8** plus one Deuce.

There are 20 cards that rank **4** through **8**. The number of 2-card combinations that contain 2 unpaired cards ranked **4** through **8** is **Comb(20, 2)** minus the 30 pairs of these ranks:

$$\frac{(20 * 19)}{(1 * 2)} = 190$$

190 − 30 = 160 is the number of 2-card combinations that will flop the Nut Low draw. The **160** is multiplied by **4** (the number of Deuces) to produce **640** 3-card combinations that will flop the nut draw to this hand:

Total Possibilities = 17,296
− WILLs = 640
WILLNOTs = 16,656

Odds of Flopping Nut Low
WILLNOTs : WILLs
16,656 : 640
Reduce
16,656 / 640 : 640 / 640
26 : 1

After the Flop

Nut Hand or Nut Draw

Because almost the entire deck (**77%**) is distributed before the Flop and is held in players' hands, it is almost certain that the Flop will have hit one or more of those hands and the presence of the nut hand and/or the nut draw is a high probability.

Only the very best starting hands offer a significant advantage over the others, because of the large number of cards held in the players' hands and the small number of cards in the Flop.

Once the Flop has hit the board, you have 7 of the 9 cards that will make your best and final hand for both Low and High. Money expectation aside, as a general rule **only continue to invest in the hand with the nuts or a reasonable draw to the nuts in either direction, preferably both.** If you don't have them, the nuts that is, there are many chances that at least one of your opponents does.

The Omaha player who receives optimal return from his game is the player who is able to continuously convince his opponents to put more money into the pot with weaker hands and lesser draws.

Money & Expectation After the Flop

After the betting on the Flop in limit games, the stakes usually double and the cost to continue in the hand increases. As each card is placed on the board the likelihood of improvement decreases. As each bet or raise is made or called the money odds get better.

In Omaha Hi-Lo the pot is often split, and many times will be split among more than 2 players.

Omaha Hi-Lo is one of the few games in poker where the absolute nut hand can wind up losing money. It is also a game where, given usually large pots, playing uni-directional hands, regardless of strength, can carry a much-reduced expectation over the same hand in non-split games.

In split games, the difference between scoop and not-scoop is dramatic. If 4 players each have $200 invested in an $800 Omaha Hi-Lo pot:

Scooper Earns $600 Profit
Low Half Earns $200 Profit
Quartered Low Earns $000 Profit

With a nut Low draw or made nut Low hand, primary considerations are:

1. Whether there will be 3 low cards on the board on the Flop or by the River

2. Whether the hand is the draw or the nuts

3. The likelihood the low will be quartered or worse

4. The chance the low will be counterfeited

5. The presence of a backup or extra low card in the hole in case of a counterfeit on the Turn or River.

With a made nut High hand or nut High draw, consider:

1. Whether or not there will be a possible Low to split the pot

2. The likelihood the hand or draw is the nuts

3. Whether the hand will make or remain the nuts

4. For certain hands, whether the board will pair

5. The likelihood that the high hand is the only high among several low hands/draws.

To optimize the opportunity with a double-nut/scoop hand or draw, consider:

1. If the hand is already made in one or both directions

2. The likelihood the hand will be made in either or both directions

3. If the hand/draw is the nuts in either or both directions

4. The odds the hand will be broken/counterfeited by the end

5. With a true double-nut hand, how to get more money into the pot.

Odds of Improvement

Once the Flop has hit the board, you have 7 of the 9 cards that will make your best and final hand. After the Flop and with 2 cards to come, regardless of the hand, any player still involved must know

♦ The odds of improvement on the next card (the Turn)

♦ The odds of improvement over the next two cards (the Turn and the River)

♦ The odds of hitting certain Runner-Runner hands.

Odds the Turn Card Will . . .

Having just one backup low card in the hole increases the probability of turning a nut Low hand from a nut Low draw from **35.6%** to **55.6%**.

With both a Low draw and a Flush draw, the player has almost a **50-50 (47%)** chance of making either the Low or the Flush and still has another card to come. A player does not need big money odds to make this hand very playable after the Flop.

With the multiple-way draw possible in Omaha Hi-Lo, the odds of improvement on the Turn can exceed **50%**. With a Low draw, Straight draw and a Flush draw, the player will turn a made hand **60%** of the time.

With certain 5-card holdings after the Flop, the odds that these holdings will improve with the laying of the Turn card are

After the Flop	Hand to Turn	Odds
Low Draw	Low	1.8 : 1
Low Draw with Backup Low Card	Low	1.1 : 1
Wheel Draw	Wheel	10.3 : 1
Low Draw or Made Low	Counterfeit Low	6.5 : 1
Low Draw & Flush Draw	Low or Flush	1.1 : 1

After the Flop	Hand to Turn	Odds
Flush Draw	Flush	4 : 1
Open Straight Draw	Straight	4.6 : 1
Gut-Shot Straight Draw	Straight	10.25 : 1
Open Straight & 4 Flush & Low Draw	Straight, Flush or Low	.67 : 1
2 Pair	Full House	10.3 : 1
Set	Full House or Quads	5.4 : 1
Made Straight or Flush	Pair the Board	4 : 1

The Calculations

With the Turn card to come, the odds calculation for the various hands is straightforward. **The only number that changes is the number of outs.**

After the Flop and before the Turn, counting your 4 hole cards, there are **45** unseen cards:

Total Possibilities = 45

Total Possibilities − WILLs (Outs) = WILLNOTs

Odds = WILLNOTs : WILLs (Outs)

Low Draw >>> Low

With 2 low cards of 2 different ranks in the hole and the same on the board, cards from 4 ranks, **4 * 4 = 16** will complete the low draw:

Total Possibilities = 45
− WILLs = 16

WILLNOTs = 29

Odds of Turning a Low
WILLNOTs : WILLs
29 : 16
Reduce
29 / 16 : 16 / 16
1.8 : 1

$$\frac{\text{WILLs}}{\text{Total Possibilities}} * 100 = \text{Probability as \%}$$

$$\frac{16}{45} * 100 = 35.6\%$$

Low Draw with Backup Low Card in the Hole >>> Low

With 3 low cards of 3 different ranks in the hole and 2 on the board, cards from the 3 remaining ranks, **3 * 4 = 12**, plus any of the **9** cards that will counterfeit any one of the hole cards, **12 + 9 = 21**, cards will complete the nut low draw:

Total Possibilities = 45
– WILLs = 21
WILLNOTs = 24

Odds of Turning a Low
WILLNOTs : WILLs
24 : 21
Reduce
24 / 21 : 21 / 21
1.1 : 1

$$\frac{\text{WILLs}}{\text{Total Possibilities}} * 100 = \text{Probability as \%}$$

$$\frac{21}{45} * 100 = 46.7\%$$

Wheel (A2345) Draw >>> Wheel

With a Wheel draw, there are only **4** cards that will make the Wheel:

Total Possibilities = 45
– WILLs = 4
WILLNOTs = 41

Odds of Turning a Wheel
WILLNOTs : WILLs
41 : 4
Reduce
41 / 4 : 4 / 4
10.3 : 1

Low Draw or Made Low Hand >>> Counterfeit Either of 2 Low Cards in the Hole

There are **6** cards that will **counterfeit** either of 2 low cards in the hole:

Total Possibilities = 45
– WILLs = 6
WILLNOTs = 39

Odds of Turning a Counterfeit to Low Card in the Hole
WILLNOTs : WILLs
39 : 6
Reduce
39 / 6 : 6 / 6
6.5 : 1

Low Draw & Flush Draw >>> Low or Flush

There are **16** cards that will complete the Low plus **9** cards that will make the flush, minus **4** cards that are already counted among those that will make the low, **16 + 9 − 4 = 21**:

Total Possibilities = 45
− WILLs = 21
WILLNOTs = 24

Odds of Turning a Low or a Flush
WILLNOTs : WILLs
24 : 21
Reduce
24 / 21 : 21 / 21
1.1 : 1

Flush Draw >>> Flush

There are **9** cards that will make the Flush on the Turn:

Total Possibilities = 45
− WILLs = 9
WILLNOTs = 36

Odds of Turning a Flush
WILLNOTs : WILLs
36 : 9
Reduce
36 / 9 : 9 / 9
4 : 1

Open Ended Straight Draw >>> Straight

There are **8** cards that will complete the Straight:

Total Possibilities = 45
- WILLs = 8
WILLNOTs = 37

Odds of Turning a Straight
WILLNOTs : WILLs
37 : 8
Reduce
37 / 8 : 8 / 8
4.6 : 1

Gut-Shot Straight Draw >>> Straight

There are **4** cards that will complete the Straight:

Total Possibilities = 45
- WILLs = 4
WILLNOTs = 41

Odds of Turning a Straight
WILLNOTs : WILLs
41 : 4
Reduce
41 / 4 : 4 / 4
10.3 : 1

Low Draw, Flush Draw & Open Ended Straight Draw >>> Low, Flush or Straight

There are **27** cards that will make a Straight, a Flush or a Low to this **great** draw:

Total Possibilities = 45
- WILLs = 27
WILLNOTs = 18

Odds of Turning a Flush or a Straight or a Low
WILLNOTs : WILLs
18 : 27
Reduce
18 / 27 : 27 / 27
.67 : 1

$$\frac{\text{WILLs}}{\text{Total Possibilities}} * 100 = \text{Probability as \%}$$

$$\frac{27}{45} * 100 = 60\%$$

2 Pair >>> Full House

There are **4** cards that will make a Full House with 2 Pair:

Total Possibilities = 45
– WILLs = 4
WILLNOTs = 41

Odds of Turning a Full Boat
WILLNOTs : WILLs
41 : 4
Reduce
41 / 4 : 4 / 4
10.3 : 1

3 of a Kind >>> Full House or Quads

There are **7** cards that will turn a Boat or Quads with a Set involving a pair in the hole:

Total Possibilities = 45
– WILLs = 7
WILLNOTs = 38

Odds of Turning a Boat or Quads
WILLNOTs : WILLs
38 : 7
Reduce
38 / 7 : 7 / 7
5.4 : 1

Turn Card Will Pair the Board

There are **9** cards that will pair one of the 3 board cards:

Total Possibilities = 45
– WILLs = 9
WILLNOTs = 36

Odds of Pairing the Board
WILLNOTs : WILLs
36 : 9
Reduce
36 / 9 : 9 / 9
4 : 1

Odds with 2 Cards to Come

With 2 cards to come there are 4 possibilities:

1. You will hit none of your outs on either card

2. You will hit one of your outs on the Turn

3. You will hit one of your outs on the River

4. You will hit one of your outs on the Turn and another on the River.

The most liberal of these calculations figures the odds of hitting an out on either the Turn or the River, or hitting an out on both the Turn and the River.

Some notable points about making certain hands with 2 cards to come are

♦ Most Low draws have an excellent chance of completion with two cards to come.

♦ With a Low draw, the Low will complete **49%** of the time.

♦ With a Low draw and a backup low card in the hole the Low will complete **69.4%** of the time.

♦ With 2 cards to come you will **only** counterfeit one of the 2 low cards in the hole **once in every four** draws.

♦ With a Flush draw and a Low draw the player will make either or both the Low or the Flush over **72%** of the time with 2 cards to come.

♦ With unpaired hole cards and an unpaired board after the Flop, the board will show a pair by the River **42%** of the time.

The odds of hitting at least one of the available outs on either or both the Turn and the River are

After the Flop	Hand After River Card	Odds
Low Draw	Low	1.03 : 1
Low Draw with Backup Low Card	Low	.4 : 1
Wheel Draw	Wheel	4.8 : 1
Low Draw or Made Low	Counterfeit Low	3 : 1

After the Flop	Hand After River Card	Odds
Flush Draw	Flush	1.75 : 1
Low Draw & Flush Draw	Low or Flush	.4 : 1
Low Draw & Flush Draw	Low & Flush	8.2 : 1
Open Straight Draw	Straight	2.1 : 1
Gut-Shot Straight Draw	Straight	4.8 : 1
2 Pair	Full House or Quads	4.6 : 1
Set	Full House or Quads	1.9 : 1
Made Straight or Flush	Pair the Board	1.4 : 1

The Calculations

After the Flop and before the Turn card has been dealt, there are 45 unseen cards. When calculating the odds with 2 cards to come, an important number is the total possible 2-card combinations that can be made from the remaining unseen 45 cards — **Comb(45, 2):**

$$\frac{(45 * 44)}{(1 * 2)} = 990$$

To find the **WILLs,** add the number of possible 2-card combinations that will contain 1 and only 1 of the available outs to the number of 2-card combinations that will contain 2 out cards.

Low Draw >>> Low with 2 Cards to Come

With a Low draw, 16 of the unseen 45 cards will make a Low. There are 6 cards among the 45 that will counterfeit one of the low cards in the hole. Of the 45 unseen cards there are **16** that will make the Low and **23** that can combine with one of those 16 to make the Low without counterfeiting one of the low cards in the hole.

16 * 23 = 368 is the number of 2-card combinations, one card of which will make the Low.

Comb(16, 2) is the number of 2-card combinations that contain 2 cards that will make the Low, including pairs:

$$\frac{(16 * 15)}{(1 * 2)} = 120$$

120 2-card combinations will contain 2 low cards and still complete the Low without counterfeiting either of the low cards in the hole.

368 + 120 = 488 possible 2-card combinations **WILL** make the Low:

Total Possibilities = 990
– WILLs = 488
WILLNOTs = 502

Odds of a Low with 2 Cards to Come
WILLNOTs : WILLs
502 : 488
Reduce
502 / 488 : 488 / 488
1.03 : 1

$$\frac{\textbf{WILLs}}{\textbf{Total Possibilities}} * 100 = \textbf{Probability as \%}$$

$$\frac{488}{990} * 100 = 49\%$$

Low Draw with Backup Low Card in the Hole >>> Low with 2 Cards to Come

With a Low draw and a backup low card in the hole, **12** of the unseen 45 cards will make a Low. A Low can also be made by counterfeiting any 1 of the 3 low cards in the hole. **12 + 9 = 21** cards among the 45 that will make the Low.

21 ∗ 24 = 504 is the number of 2-card combinations, one card of which will make the low.

Comb(21, 2) is the number of 2-card combinations that contain 2 cards either of which will make a Low, including pairs. **To make this Low, none of those 2-card combinations may counterfeit more than one rank of the cards in the hole.**

27 2-card combinations each of which contain a card that counterfeits a different rank in the hole, must be subtracted from **Comb(21, 2)** to find the number of 2-card combinations that will contain 2 low cards and still successfully make a Low to this hand:

$$\frac{(21 * 20)}{(1 * 2)} = 210$$

$$210 - 27 = 183$$

504 + 183 = 687 possible 2-card combinations **WILL** make the low:

$$\text{Total Possibilities} = 990$$
$$- \text{WILLs} = 687$$
$$\text{WILLNOTs} = 303$$

Odds of a Low with 2 Cards to Come
WILLNOTs : WILLs
303 : 687
Reduce
303 / 687 : 687 / 687
.4 : 1

$$\frac{\text{WILLs}}{\text{Total Possibilities}} * 100 = \text{Probability as \%}$$

$$\frac{687}{990} * 100 = 69\%$$

Wheel Draw >>> Wheel with 2 Cards to Come

With a Wheel draw, only 4 of the unseen 45 cards will make the Wheel. As long as this last card to the Wheel is one of the 2 final board cards, it will not affect the value of the hand if 1 of the 2 last cards counterfeits either a board card or a hole card. Of the 45 unseen cards, **4** will make the Wheel and the remaining **41** can combine with one of these 4 to make the Wheel.

4 * 41 = 164 is the number of 2-card combinations, one card of which will make the Wheel.

Comb(4, 2) is the number of 2-card combinations that contain 2 cards that will make the Low, including pairs:

$$\frac{(4 * 3)}{(1 * 2)} = 6$$

6 2-card combinations will contain 2 Wheel cards and still complete the Low regardless of counterfeiting either of the low cards in the hole.

164 + 6 = 170 possible 2-card combinations **WILL** make the Wheel:

Total Possibilities = 990
– WILLs = 170
WILLNOTs = 820

Odds of a Wheel with 2 Cards to Come
WILLNOTs : WILLs
820 : 170
Reduce
820 / 170 : 170 / 170
4.8 : 1

Low Draw >>> Counterfeit One or More Low Cards in the Hole with 2 Cards to Come

With a Low draw there are 6 cards that can counterfeit either or both low cards in the hole. **6** of the unseen 45 cards will counterfeit the Low and **39** will not.

6 * 39 = 234 is the number of 2-card combinations, one card of which will counterfeit one of the low cards in the hole.

Comb(6, 2) is the number of 2-card combinations in which both cards match one or both of the low cards in the hole:

$$\frac{(6 * 5)}{(1 * 2)} = 15$$

234 + 15 = 249 possible 2-card combinations **WILL** counterfeit at least one of the 2 low cards in the hole:

Total Possibilities = 990
– WILLs = 249
WILLNOTs = 741

**Odds of Counterfeiting
At Least 1 of 2 Low Cards in the Hole
WILLNOTs : WILLs
741 : 249
Reduce
741 / 249 : 249 / 249
3 : 1**

Flush Draw >>> Flush with 2 Cards to Come

With a Flush draw and 45 unseen cards, **9** cards will make the Flush and **36** will not.

9 ∗ 36 = 324 is the number of 2-card combinations, one card of which will make the Flush.

Comb(9, 2) is the number of 2-card combinations that contain 2 cards either of which make the Flush:

$$\frac{(9 * 8)}{(1 * 2)} = 36$$

324 + 36 = 360 possible 2-card combinations **WILL** make the Flush:

**Total Possibilities = 990
− WILLs = 360
WILLNOTs = 630**

**Odds of Making a Flush with 2 Cards to Come
WILLNOTs : WILLs
630 : 360
Reduce
630 / 360 : 360 / 360
1.75 : 1**

Flush Draw & Low Draw >>> Flush or Low

With a Flush draw and a Low draw there are **16** low cards that will complete the Low. 4 of these are of the same suit as the draw, so there are **16 + 5 = 21** cards that will make either the Flush or the Low. **21** of the unseen 45 cards will make the Flush or the low and **24** will not.

21 * 24 = 504 is the number of 2-card combinations, one card of which will make either the Flush or the low.

Comb(21, 2) is the number of 2-card combinations that contain 2 cards both of which can make either the Flush or the low:

$$\frac{(21 * 20)}{(1 * 2)} = 210$$

504 + 210 = 714 possible 2-card combinations **WILL** make the Flush:

Total Possibilities = 990
– WILLs = 714
WILLNOTs = 276

Odds of a Flush or a Low with 2 Cards to Come
WILLNOTs : WILLs
276 : 714
Reduce
276 / 714 : 714 / 714
.4 : 1

$$\frac{\textbf{WILLs}}{\textbf{Total Possibilities}} * 100 = \textbf{Probability as \%}$$

$$\frac{714}{990} * 100 = 72\%$$

Low Draw & Flush Draw >>> Flush & Low

With a Flush draw and a Low draw, there are **12** cards that will make the Low without making the Flush, and **9** cards that will make the Flush without making the Low.

12 * 9 = 108 is the number of 2-card combinations that will make both the Flush and the Low:

Total Possibilities = 990
− WILLs = 108
WILLNOTs = 882

Odds of Making BOTH a Flush and a Low
with 2 Cards to Come
WILLNOTs : WILLs
882 : 108
Reduce
882 / 108 : 108 / 108
8.2 : 1

Open Ended Straight Draw >>> Straight

With a Straight draw and 45 unseen cards, there are **8** cards that will make the Straight and **37** will not.

Therefore, **8 * 37 = 296** is the number of 2-card combinations, one card of which will make the Straight.

Comb(8, 2) is the number of 2-card combinations that contain 2 cards either of which will make the Straight:

$$\frac{(8 * 7)}{(1 * 2)} = 28$$

296 + 28 = 324 possible 2-card combinations **WILL** make the Straight:

Total Possibilities = 990
– WILLs = 324
WILLNOTs = 666

Odds of Making a Flush with 2 Cards to Come
WILLNOTs : WILLs
666 : 324
Reduce
666 / 324 : 324 / 324
2.1 : 1

Gut-Shot Straight Draw >>> Straight with 2 Cards to Come

With a gut-shot Straight draw and 45 unseen cards, there are **4** cards that will make the Straight and **41** will not.

4 * 41 = 164 is the number of 2-card combinations, one card of which will make the Straight.

Comb(4, 2) is the number of 2-card combinations that contain 2 cards either of which will make the Straight:

$$\frac{(4 * 3)}{(1 * 2)} = 6$$

164 + 6 = 170 possible 2-card combinations that **WILL** make the Straight:

Total Possibilities = 990
– WILLs = 170
WILLNOTs = 820

Odds of Making a Gut-Shot Straight with 2 Cards to Come
WILLNOTs : WILLs
820 : 170
Reduce
820 / 170 : 170 / 170
4.8 : 1

2 Pair >>> Full House or Quads

With 2 pair, assuming no pairs in the hole, there are 4 cards, any one of which will make a Full House or Quads. **4** of the unseen 45 cards will make the hand and **41** will not.

4 * 41 = 164 is the number of 2-card combinations, one card of which will make the Boat.

Comb(4, 2) is the number of 2-card combinations that contain 2 cards either of which make the Boat or Quads:

$$\frac{(4 * 3)}{(1 * 2)} = 6$$

In addition there are **8** pairs that can also make the Boat.

164 + 6 + 8 = 178 possible 2-card combinations that **WILL** make the Boat or Quads:

Total Possibilities = 990
− WILLs = 178
WILLNOTs = 812

Odds of making a Full Boat or Quads with 2 Cards to Come
WILLNOTs : WILLs
812 : 178

Reduce
812 / 178 : 178 / 178
4.6 : 1

Set >>> Full House or Quads

With a Set involving a pair in the hole and 45 unseen cards, there are **7** cards that will make a Boat or Quads and **38** will not.

7 * 38 = 266 is the number of 2-card combinations, one card of which will make the Boat or Quads.

Comb(7, 2) is the number of 2-card combinations that contain 2 cards, either of which will make the Boat or Quads:

$$\frac{(7 * 6)}{(1 * 2)} = 21$$

In addition, there are **54** different pairs that can come in the last 2 cards to make the Full House.

266 + 21 + 54 = 341 possible 2-card combinations that **WILL** make the Boat or Quads:

Total Possibilities = 990
– WILLs = 341
WILLNOTs = 649

**Odds of making a Full Boat or Quads
with 2 Cards to Come**
WILLNOTs : WILLs
649 : 341
Reduce
649 / 341 : 341 / 341
1.9 : 1

Unpaired Hole Cards >>> Board Will Show a Pair

There are **9** cards than can pair one of the three board cards and **36** will not.

9 * 36 = 324 is the number of 2-card combinations, one card of which will pair the board.

Comb(9, 2) is the number of 2-card combinations that contain 2 cards either of which will pair the board:

$$\frac{(9 * 8)}{(1 * 2)} = 36$$

In addition there are **57** pairs that can also come to create a pair on the board.

324 + 36 + 57 = 417 possible 2-card combinations that **WILL** pair the board:

Total Possibilities = 990

– WILLs = 417

WILLNOTs = 573

Odds the Board Will Pair with 2 Cards to Come

WILLNOTs : WILLs

573 : 417

Reduce

573 / 417 : 417 / 417

1.4 : 1

$$\frac{\textbf{WILLs}}{\textbf{Total Possibilities}} * 100 = \text{Probability as \%}$$

$$\frac{417}{990} * 100 = 42\%$$

Runner – Runner

Remember — runner-runners don't complete very often and in most cases require extraordinary money odds to be played profitably. **No one card will do it.**

The Calculations

The 2 most common ways to calculate the odds of hitting a runner-runner after the Flop are

1. Calculate the number of 2-card combinations that **WILL** do the job and compare that to the total possible 2-card combinations to find the number of 2-card combinations that **WILLNOT** complete the hand.

2. Multiply the probability of the first runner by the probability of the second runner and then convert that probability to odds.

This section demonstrates both approaches.

For the combination, we need to know the total possible 2-card combinations from the remaining 45 unseen cards:

$$\frac{(45 * 44)}{(1 * 2)} = 990$$

Below are the odds and calculations:

After The Flop	Runner – Runner	Odds
Back Door Low Draw	**Low**	**5.2 : 1**
Back Door Wheel Draw	**Wheel**	**21.5 : 1**

After the Flop	Runner – Runner	Odds
Back Door Flush Draw	**Flush**	**26.5 : 1**
Back Door Open Ended Straight Draw	**Straight**	**19.6 : 1**
4 Unpaired Hole Cards	**Set**	**81.5 : 1**
4 Unpaired Hole Cards	**Pair 2 Hole Cards or Set**	**66.56 : 1**
2 Pair in the Hole	**Quads**	**494 : 1**
Open Back Door Straight Flush Draw	**Straight Flush**	**329 : 1**

Back Door Low Draw >>> Runner-Runner Low

With 2 low cards in the hole and 1 usable low card on the board, even if the board low card is paired, there are **20** cards of 5 ranks, any unpaired 2 of which will make the runner-runner Low. **Comb(20, 2)** minus the **30** pairs possible among the cards from these 5 ranks is the number of 2-card combinations that will make the runner-runner Low:

$$\frac{(20 * 19)}{(1 * 2)} = 190$$

190 – 30 = 160 possible 2-card combinations that **WILL** make a runner-runner Low:

Total Possibilities = 990
– WILLs = 160
WILLNOTs = 830

Odds of a Runner-Runner Low
WILLNOTs : WILLs
830 : 160
Reduce

830 / 160 : 160 / 160

5.2 : 1

Back Door Wheel Draw >>> Runner-Runner Wheel

With 2 wheel cards in the hole and 1 usable wheel card on the board, even if the board wheel card is paired, there are **8** cards of 2 ranks, any unpaired 2 of which will make the runner-runner Wheel. **Comb(8, 2)** minus the **12** pairs possible among the cards from these 2 ranks, is the number of 2-card combinations that will make the runner-runner Wheel

$$\frac{(8 * 7)}{(1 * 2)} = 28$$

28 − 12 = 16 possible 2-card combinations that **WILL** make a runner-runner Wheel:

Total Possibilities = 990

− WILLs = 16

WILLNOTs = 974

Odds of a Runner-Runner Wheel

WILLNOTs : WILLs

974 : 16

Reduce

974 / 16 : 16 / 16

60.9 : 1

Back Door Flush Draw >>> Runner-Runner Flush

With 2 suited cards in the hole and 2 suited cards on the board, there are **9** cards any 2 of which will make the runner-runner Flush. **Comb(9, 2)** is the number of 2-card combinations that will make the runner-runner Flush:

$$\frac{(9 * 8)}{(1 * 2)} = 36$$

36 possible 2-card combinations that **WILL** make a runner-runner Flush:

Total Possibilities = 990
– WILLs = 36
WILLNOTs = 954

Odds of a Runner-Runner Flush
WILLNOTs : WILLs
954 : 36
Reduce
954 / 36 : 36 / 36
26.5 : 1

Open Backdoor 3-Straight >>> Runner-Runner Straight

With 3 connected cards above **2** and below **K** in rank, and with two of those cards in the hole, there are **3** sets of 2-card combinations that will make the Straight:

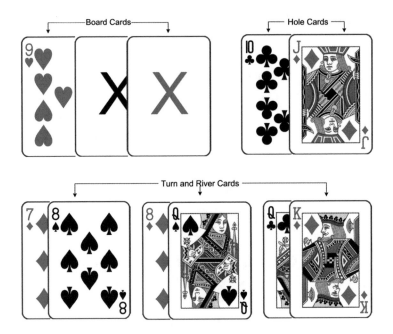

For each of these 3 sets there are **4 * 4 = 16** possible combinations. There is a total of **3 * 16 = 48** 2-card combinations that will make the Straight shown:

Total Possibilities = 990
– WILLs = 48
WILLNOTs = 942

Odds of a Runner-Runner Straight
WILLNOTs : WILLs
942 : 48
Reduce
942 / 48 : 48 / 48
19.6 : 1

4 Unpaired Cards in the Hole >>> Runner-Runner Set

With 4 unpaired cards in the hole, to make a runner-runner Set the pair that matches one of the **4** hole cards will have to hit on the Turn and another on the River. There are **3** pairs outstanding for each of the 4 ranks in the hole. **4 ∗ 3 = 12** pairs will make a Set to one of the hole cards:

Total Possibilities = 990
– WILLs = 12
WILLNOTs = 978

Odds of a Runner-Runner Set
WILLNOTs : WILLs
978 : 12
Reduce
978 / 12 : 12 / 12
81.5 : 1

4 Unpaired Hole Cards >>> Runner-Runner 2 Pair or Set

With 4 unpaired cards in the hole there are **12** cards, any 2 of which will make a runner-runner **2** pair or a Set to the cards in the hole — **Comb(12, 2)**:

$$\frac{(12 * 11)}{(1 * 2)} = 66$$

66 possible 2-card combinations that **WILL** make a runner-runner 2 pair or a Set to 4 unpaired cards in the hole:

Total Possibilities = 990
– WILLs = 66
WILLNOTs = 924

Odds of a Runner-Runner 2 Pair or Set
WILLNOTs : WILLs
924 : 66
Reduce
924 / 66 : 66 / 66
14 : 1

2 Pair in the Hole >>> Runner-Runner Quads

With 2 pair in the hole, there are only **2** pairs or 2 2-card combinations that **WILL** make runner-runner Quads to this hand:

Total Possibilities = 990
– WILLs = 2
WILLNOTs = 988

Odds of a Runner-Runner Quads
WILLNOTs : WILLs
988 : 2
Reduce
988 / 2 : 2 / 2
494 : 1

Open Back Door 3-Straight Flush >>> Runner-Runner Straight Flush

With 2 suited and connected cards above **2** and below **K** in rank on the board, and 1 in the hole, there are **3** 2-card combinations that **WILL** make a Straight Flush

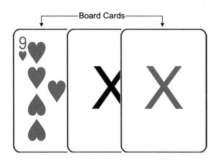

Board Cards

Hole Cards

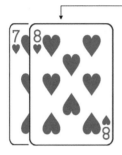

Turn and River Cards

Total Possibilities = 990
– WILLs = 3
WILLNOTs = 987

Odds of a Runner-Runner Straight Flush
WILLNOTs : WILLs
987 : 3
Reduce
987 / 3 : 3 / 3
329 : 1

Before the Last Card

After the Turn card has hit the board, you have only 1 more card to come and only 1 more chance to improve your hand.

With 1 card to come there are still 2 rounds of betting left and in most limit games those rounds are at a higher stake than the first two.

With the higher cost to continue and the hand mostly made, players should have a precise strategy in order to remain involved.

At this juncture players have 8 out of the 9 cards that will be used to make both Low and High hands. The player should have a very good idea of his chances:

1. That he is **drawing dead** in either or both directions

2. That he has the nuts or will make the nuts for either or both the High or the Low

3. That he will be quartered or counterfeited for the Low.

Odds of Improvement With Only 1 Card to Come

With only 1 card to come, some noteworthy points from the table below include:

1. With a Low draw, you will make the Low roughly **36%** of the time

2. Having just 1 backup low card in the hole raises your probability of making the Low with only 1 card to come from **36%** to **48%**

3. While some of the odds seem quite favorable, a player strong in only 1 direction should remember that most of the time he will be playing for only **half the pot.**

Here is a table that shows the odds of improving certain common drawing hands on the last card:

Hand – Draw	Make on Last Card	Odds
Low Draw	Low	1.75 : 1
Low Draw with Backup Low Card in the Hole	Low	1.1 : 1
Low Draw & Flush Draw	Low or Flush	1.1 : 1
2 Pair	Full House	10 : 1
Open Ended Straight Draw	Straight	4.5 : 1
Gut-Shot Straight Draw	Straight	10 : 1
Flush Draw	Flush	3.9 : 1
Flush Draw & Open Straight Draw	Straight or Flush	1.9 : 1
1 Pair & Flush Draw	Set or Flush	3 : 1
Open Straight Flush Draw	Straight Flush	21 : 1
4 Unpaired Hole Cards	Pair Any Hole Card	2.7 : 1

The Calculations

With one card to come, the odds calculation for the various hands is straightforward. **The only number that changes is the number of outs.**

After the Turn card has hit the board, there are 4 cards in the hole, 4 on the board and **44** unseen cards.

Total Possibilities = 44

Total Possibilities − WILLs (Outs) = WILLNOTs

Odds = WILLNOTs : WILLs (Outs)

Low Draw >>> Low

With a Low draw there are **16** cards from 4 ranks that can come on the River to make the Low:

Total Possibilities = 44
− WILLs = 16
WILLNOTs = 28

Odds of Making a Low
WILLNOTs : WILLs
28 : 16
Reduce
28 / 16 : 16 / 16
1.75 : 1

$$\frac{\text{WILLs}}{\text{Total Possibilities}} * 100 = \text{Probability as \%}$$

$$\frac{16}{44} * 100 = 36\%$$

Low Draw with Backup Low Card in the Hole >>> Low

With a Low draw and a backup low card in the hole, there are **21** cards that can come on the River to make the Low:

Total Possibilities = 44
– WILLs = 21
WILLNOTs = 23

Odds of Making a Low
with Backup Low Card in the Hole
WILLNOTs : WILLs
23 : 21
Reduce
23 / 21 : 21 / 21
1.1 : 1

$$\frac{\textbf{WILLs}}{\textbf{Total Possibilities}} * 100 = \text{Probability as \%}$$

$$\frac{21}{44} * 100 = 48\%$$

Low Draw & Flush Draw >>> Low or Flush

With a Low draw and a Flush draw, there are **21** cards that can come on the River to make the Flush or the Low:

Total Possibilities = 44
– WILLs = 21
WILLNOTs = 23

Odds of Making a Flush or a Low
WILLNOTs : WILLs
23 : 21

Reduce
23 / 21 : 21 / 21
1.1 : 1

2 Pair >>> Full House

With 2 pair there are only **4** cards that can come on the River to make the Full House:

Total Possibilities = 44
– WILLs = 4
WILLNOTs = 40

Odds of Making a Boat with 2 Pair
WILLNOTs : WILLs
40 : 4
Reduce
40 / 4 : 4 / 4
10 : 1

Open Ended Straight Draw >>> Straight

With an open ended Straight draw there are **8** cards that can come on the River to make the Straight:

Total Possibilities = 44
– WILLs = 8
WILLNOTs = 36

Odds of Making a Straight
WILLNOTs : WILLs
36 : 8
Reduce
36 / 8 : 8 / 8
4.5 : 1

Gut-Shot Straight Draw >>> Straight

With a gut-shot Straight draw there are only **4** cards that can come on the River to make the Straight:

Total Possibilities = 44
– WILLs = 4
WILLNOTs = 40

Odds of Making a Straight
WILLNOTs : WILLs
40 : 4
Reduce
40 / 4 : 4 / 4
10 : 1

Flush Draw >>> Flush

With a Flush draw there are **9** cards that can come on the River to make the Flush:

Total Possibilities = 44
– WILLs = 9
WILLNOTs = 35

Odds of Making a Flush
WILLNOTs : WILLs
35 : 9
Reduce
35 / 9 : 9 / 9
3.9 : 1

Open Ended Straight Draw & Flush Draw >>> Flush or Straight

With an open-ended Straight draw and a Flush draw there are **15** cards that can come on the River to make either the Flush or the Straight:

Total Possibilities = 44
– WILLs = 15
WILLNOTs = 29

Odds of Making a Straight or a Flush
WILLNOTs : WILLs
29 : 15
Reduce
29 / 15 : 15 / 15
1.9 : 1

1 Pair & Flush Draw >>> Flush or Set

With a Pair and a Flush draw there are **11** cards that can come on the River to make the Flush or a Set on the River:

Total Possibilities = 44
– WILLs = 11
WILLNOTs = 33

Odds of Making a Flush or a Set
WILLNOTs : WILLs
33 : 11
Reduce
33 / 11 : 11 / 11
3 : 1

Open Ended Straight Flush Draw >>> Straight Flush

With an open ended Straight Flush draw there are only **2** cards that can come on the River to make the Flush:

Total Possibilities = 44
– WILLs = 2
WILLNOTs = 42

Odds of Making a Straight Flush
WILLNOTs : WILLs
42 : 2
Reduce
42 / 2 : 2 / 2
21 : 1

4 Unpaired Hole Cards >>> Pair a Hole Card

With 4 unpaired hole cards there are **12** cards that will pair one of them on the River:

Total Possibilities = 44
– WILLs = 12
WILLNOTs = 32

Odds of Making a Pair on the River
WILLNOTs : WILLs
32 : 12
Reduce
32 / 12 : 12 / 12
2.7 : 1

The River Bet

All the cards are out. There is no chance of improving your hand. You can win a piece of the pot in one of two ways:

1. By having the **best hand** in a showdown

2. By using **strategy** to cause your opponent(s) to fold their cards.

Your **Total Odds** of winning the pot are the odds that your hand will win in a showdown **PLUS** the odds that you can manipulate your opponent(s) into giving up their stake in the pot.

In Omaha Hi-Lo, however, because of the high probability of a call from either or both a strong Low or a strong High hand, moves are not as useful as they can be in Hold'em.

It is at this point in a hand of Omaha Hi-Lo where the concepts of **Value-Bet** and even more so — **Value-Call** — are most plainly demonstrated.

By this stage, three of the four rounds of betting have been completed and the pot can be monstrous. Pots are often large enough to offer positive money odds to a variety of seemingly risky propositions.

Value-Bluffs might not be effective at this point. But marginal strength in **both** directions could easily warrant a **Value-Call.**
